'Read Barry's book. No one has ̶ ̶ ̶ ̶ ̶ ̶ ̶ ̶ ̶ ̶ ̶ ̶
longer, harder, and deeper about this country.'

PHILLIP ADAMS AO

'The hope for this book is that the young people who read it
will respond to its scientifically sound and brilliant analyses
of how our society needs to change post Covid. Their health
and wellbeing as well as that of the planet depends on
citizens being informed and challenging our undemocratic
political culture.'

PROFESSOR FIONA STANLEY AC

'The author of this book is a genius. He irritates the hell
out of people of all political loyalties. He reads virtually
everything that matters. In these pages we, his readers,
are the beneficiaries. Forty years after his masterpiece,
Sleepers, Wake!, he tackles the challenges of a new
age: the digital world, climate change, COVID-19, and
widespread political disillusionment. If any author can
offer us thoughtful directions for what is to be done,
it is Barry Jones.'

MICHAEL KIRBY AC CMG

'If anyone has written a more precise distillation of the
current issues and implications of climate change, I haven't
read or heard of it. It would be a sorrowful joy to see
our politicians try to answer Barry Jones' piercing list of
questions on climate change. They would doubtless miss
the irony embedded in every one.'

GEOFF COUSINS

'Some of the issues that Barry Jones addressed almost
40 years back in *Sleepers, Wake!* persist, while new
problems have emerged. Now, in an even more challenged
world, he asks the single most important question:
"What's to be done?"'

PROFESSOR PETER DOHERTY AC

WHAT IS TO BE DONE

Barry Jones is a former Labor member of the Victorian and Commonwealth parliaments who led the campaign to abolish the death penalty, and became Australia's longest-serving minister for science, from 1983 to 1990. His books include *Sleepers, Wake!*, *A Thinking Reed*, *Dictionary of World Biography*, and *The Shock of Recognition*. The only person to be elected as a Fellow of four of Australia's five learned academies, he became a 'living national treasure' in 1997, and was made a Companion in the Order of Australia, the nation's highest award, in 2014.

To Rachel—the present, and Emlyn
(aged six)—the future.
He has concentrated my mind
wonderfully.

WHAT IS TO BE DONE

POLITICAL ENGAGEMENT
AND SAVING THE PLANET

Barry Jones

SCRIBE
Melbourne • London

Scribe Publications
2 John St, Clerkenwell, London, WC1N 2ES, United Kingdom
18–20 Edward St, Brunswick, Victoria 3056, Australia
3754 Pleasant Ave, Suite 100, Minneapolis, Minnesota 55409, USA

Published by Scribe 2020

Typeset in Adobe Garamond Pro by J&M Typesetting

Printed and bound in the UK by CPI Group (UK) Ltd, Croydon CR0 4YY

Scribe Publications is committed to the sustainable use of natural resources and the use of paper products made responsibly from those resources.

9781913348038 (UK edition)
9781950354320 (US edition)
9781925849912 (Australian edition)
9781925938357 (ebook)

Catalogue records for this book are available from the National Library of Australia and the British Library.

scribepublications.co.uk
scribepublications.com
scribepublications.com.au

Man is but a reed, the feeblest in nature, but he is a thinking reed. There is no need for the whole universe to take up arms to crush him. A vapour or a drop of water is enough to kill him. But even if the universe were to crush him, man would still be nobler than his killer, for he knows that he is dying and that the universe has the advantage over him. The universe knows nothing of this.

Thus all our dignity consists in thought. It is on thought that we must depend for our recovery, not on space or time, which we could never fill. Let us then strive to think well; that is the basic principle of morality.

— Blaise Pascal, *Pensées*

Contents

Preface

I want to be there when everyone suddenly understands what it
has all been for.

– FYODOR DOSTOEVSKY, *THE BROTHERS KARAMAZOV* (1881)

Sleepers, Wake!: technology and the future of work, first published in
1982, was an attempt to describe the impact of technological change,
especially the information revolution, on employment, industry,
education, and training. I aspired to make a grand synthesis, linking
politics, history, economics, science, technology, education, the
concept of time-use value, psychology, and information theory.

What Is to Be Done is not an update or a revision—too much
has changed since 1982—but a sequel, addressing the massive global
changes that have occurred since.

A post-industrial work force, the digital revolution, universal
access to higher education, and the emergence of a 'third age' after
retirement were all novel concepts in 1982, and even after the last
revision of the book in 1995. Now we take them for granted, but
they did not develop as I had hoped.

There are nearly 5 billion users of the Internet, more than
60 per cent of the world's population. Our handheld devices have
more capacity than the mainframe computers used in the 1969 moon
landing, giving us instant access to the world's intellectual resources.

As well, the United States, the United Kingdom, and Australia
have by far the largest cohort of tertiary-qualified citizens in their

history. This ought to provide us with an unparalleled capacity to understand the world's greatest problems—climate change, the refugee crisis, degradation of the environment, poverty, pandemics, the exploitation of women and children, terrorism—with an informed population and inspirational leadership.

However, in the digital age, far from exploring the universal and the long term, both mainstream and social media emphasise the personal, or the tribal, in the short term. Opinion is preferenced over evidence, and feeling over rationality, while science and free enquiry are rejected or discounted. Empathy, the common good, and preserving the planet have low priority.

The planet, notoriously, has no vote.

Homo sapiens has morphed into *Homo economicus*, because all our politics revolves around production and consumption.

What is sometimes called 'the Enlightenment project' has come under sustained attack in the United States, much of Europe, and, to some extent, Australia. Instead, we see a retreat from reason; the rejection of facts and expertise; the rise of populism, snarling nationalisms, tribalism, and conspiracy theories; a fundamentalist revival and hostility to science; a failure of ethical leadership; deepening corruption of democratic processes; profound neglect of the climate-change imperative; and the triumph of vested interests. All are existential threats to civilisation's advancement and the welfare of humanity here and elsewhere.

The greatest threat to liberal democracy and Enlightenment values has not been external—from ISIS/the Taliban/al-Qaeda, China, Russia, or even pandemics—but internal and self-inflicted.

The four horsemen of the apocalypse that threaten humanity are:

- population growth exacerbated by per capita resource use;
- climate change;

- pandemics; and
- racism and state violence.

All four are inextricably linked. The pressure on resources, compounded by the threat of climate change, has been a major factor in tribal and racial conflicts over access to water and arable land. Meanwhile, millions of refugees are blamed for seeking security for their families, inequality grows exponentially, pandemics have devastating impacts not only for the aged, but on racial minorities who are stressed by insecurity, leading in turn to violent over-reactions by the custodians of law and order.

Only racism and state violence can be tackled at a national level.

Donald Trump's election as president of the United States in 2016 was a turning point in modern history. There are far more references to him in this book than to any other person. He has transformed politics beyond recognition—and he has imitators, all over the world.

In the US, the UK, Australia, France, and other European countries, there has been a striking shift in political allegiances, centred on the 'culture wars'. The cleavage is not on economic issues, but on race, gender, religion, and attitudes to modernity and globalism. People in the higher socioeconomic levels are becoming more progressive, eager to embrace change and take risks; those in the lower levels are more conservative, anxious about change, and risk-averse, seeing themselves as potential victims.

When I began writing *What Is to Be Done*, the book was to be structured around climate change/global warming and the world's failure to act. Back in 1982, I was well aware of this threat, and can claim to have been the first Australian politician to have grasped its significance. However, I did not discuss it in *Sleepers, Wake!*

But as I worked on this manuscript in 2020, new issues kept forcing me to rethink and recalibrate, and they were all inter-related:

the coronavirus pandemic, growing inequality, misogyny, the appeal of fundamentalism, the breakdown of constitutional guarantees, state violence, state secrecy, the environmental stress caused by urbanisation and population growth, and gross increases in xenophobia, racism, and intolerance, culminating in the worldwide Black Lives Matter demonstrations. And the economic and social impact of climate change would exacerbate all these problems.

Australia can be outstanding in confronting crises, such as HIV/AIDS, the Global Financial Crisis, COVID-19, and most natural disasters. But its performance was patchy during the long, horrific bushfire season of 2019–20, and it has been woeful in failing to address climate change and transitioning to a post-carbon economy. COVID-19 demonstrated how well the federation could work, and it remains to be seen if this can be maintained in the post-pandemic era.

The better angels of our nature have been well hidden in our politics, with our part-time parliaments, the absence of serious debate, revolving-door prime ministerships, venality, vindictiveness, mediocrity, secrecy, and the influence of vested interests.

Science (medical science excepted) is on the retreat, and the universities are under attack.

I have often used Australian examples to illustrate my arguments, because it is the country I know best, and the evidence is at hand. Nevertheless, my analysis is generally applicable to all technologically dependent societies, especially the United States, Great Britain, France, Germany, Canada, and New Zealand.

Citizens everywhere must engage in the great issues, and work together to master evidence and develop our capacity to define, debate, and decide. Without it, our fellow humans will be staring into an abyss.

Despite all this, we have to be optimistic that we will have the wisdom, courage, and skill to save the planet—and ourselves. It's the only way to go.

CHAPTER ONE

Where We Begin: Sleepers, Wake! in 1982

Sleepers, Wake!, subtitled *technology and the future of work*, was published by Oxford University Press in March 1982. I was then a Labor member of the Australian House of Representatives, a shadow minister, and soon to become minister for science.

From 1983 to 1990 I was Australia's longest-serving minister for science—partly, I think, because nobody else wanted the job. I have had a lifelong interest in science/research and its implications for public policy and politics generally. After serving as a member of the Victorian and Commonwealth parliaments for a total of 26 years, and as a minister for seven, I left politics with a profound sense of frustration and unease.

Political colleagues saw me as too individual and idiosyncratic (that is, 'weird'), not a team player, and totally lacking in the killer instinct, while many in the academic community might have seen me as too political, even too populist.

In politics, my timing was appalling. I kept raising issues long before their significance was widely recognised. That made me, not a prophet, but an isolated nerd.

I can claim to have put six issues on the national agenda, but started talking about them ten, fifteen, or twenty years before many of my political colleagues were ready to listen. However, there were eager audiences in schools, tertiary institutions, and in the professions, and some interest in the press and media.

I was the first Australian politician to raise public awareness of global warming/climate change; our emerging transformation from an 'industrial' to a 'post-industrial' economy/society; the looming information revolution and transition to a digital society/economy; biotechnology; 'the Third Age'; and the importance of preserving Antarctica as a wilderness.

I predicted—accurately, as it turned out—Australia's transition from an industrial economy, producing things that hurt if you dropped them on your foot, to a post-industrial (or service) economy, and then further to a knowledge (or information) society.

In politics, timing is (almost) everything. The best time to raise an issue is about ten minutes before its importance becomes blindingly obvious to the community generally.

However, there were some consolations.

I am the only person, so far, to have been elected as a Fellow of four of Australia's five learned academies—Science, Humanities, Technological Sciences and Engineering, and Social Sciences. Awarded an AC (Companion of the Order of Australia) in 2014, I was gratified that it had not been posthumous.

Where *Sleepers* came from

I was born—just—in the first third of the twentieth century, a period that began on 1 January 1901 and ended on 30 April 1934.

The inauguration of the Commonwealth of Australia and the deaths of Queen Victoria and Giuseppe Verdi all occurred in January 1901. Then followed the first aircraft, the mass production of motor vehicles, the beginnings of feature films, radio, even primitive television, Einstein's theory of relativity and E = mc², *The Rite of Spring*, *Ulysses*, World War I, the Russian Revolution, the Great Depression, the dictatorships of Lenin, Stalin, Mussolini, and Hitler, Roosevelt's 'New Deal', and Gandhi's campaigns for Indian freedom.

The year 1913 had been marked by an extraordinary mood of optimism about a golden age that might go on forever. That mood came crashing down twelve months later, and did not fully recover for about 40 years.

I am old enough to remember the end of the Great Depression, blackouts, gas masks, rationing, and slit trenches during World War II, night-soil collection in many suburbs, milk bottles filled each morning, blocks of ice delivered by a man with a horse and cart, and eleven mail deliveries each week.

From early childhood I had a stark vision of how technological change could enlarge human capacity or threaten it, so that workers become slaves of the machine—something that Karl Marx had predicted in 1858.

Most of the themes in *Sleepers* had been in my head from childhood and, as usual, the main sources were films and books, and, later, personal contacts. Charles Chaplin's *Modern Times* (1936) had a great impact on me when I first saw it at the age of eight or nine, with the powerful imagery of the nameless worker as a mere bolt-tightener for a machine.

Later, I saw René Clair's *À nous la liberté* (1931), an anarchic comedy about the dehumanisation of industrial workers. (Clair accused Chaplin of plagiarism when *Modern Times* appeared.) And I read about the play *RUR* (*Rossum's Universal Robots*) by the Czech dramatist Karel Čapek, who coined the word 'robot'.

I never became a science-fiction obsessive, but I read Jules Verne and H.G. Wells carefully.

Jules Verne (1828–1905), writing in the 1860s and 1870s, gave detailed descriptions of submarines, motion pictures, television, helicopters, space travel, storage batteries to replace coal, and a variety of household devices.

H.G. Wells (1866–1946), the English novelist, historian, and scientific prophet, once (but no longer) a household name, was an early influence. In 1899, he had written a novel *When the Sleeper Wakes*, revised and republished in 1910 as *The Sleeper Awakes*. I was unaware of either title when I planned my book.

The film *Things to Come* (1936), based on a novel by Wells, produced by Alexander Korda and stylishly directed by William Cameron Menzies, with evocative music by Arthur Bliss, was set in the year 2036. It predicted the destruction of Britain by war, a descent into chaos, and reconstruction by a scientific elite. My selective reading confirmed that many writers had made plausible descriptions of future technological developments and their impact on life, work, and society. Wells, before becoming depressive himself, generally presented a cheerful, utopian vision of man as an optimistic systems builder, determined to solve the problems of humanity.

I became preoccupied with the question of whether people should be performing repetitive, boring, dangerous, exhausting tasks when machines could do them better, and with the problem of trying to balance liberty and security.

Aldous Huxley (1894–1963), part of the famous English scientific and literary clan, including Thomas, Julian, and Andrew, was a witty and perceptive novelist and essayist of exceptional range. His *Brave New World* (1932), a dystopian fable, holds up well after almost 90 years.

Huxley predicted a totalitarian welfare state, without war, poverty, or crime, highlighted by the production of test-tube babies; the decline of the family; designer drugs; the prolongation of youth; the use of subsistence agriculture, not for production but to absorb labour; the saturation of time by media; and the development of a leisure society. In Australia, *Brave New World* was banned as obscene until 1937.

Brave New World was set in a benevolent dictatorship. Huxley wrote before Hitler came to power, Stalin initiated major purges, and World War II plunged the world into mass extermination and other brutalities.

George Orwell (1903–50) wrote *Nineteen Eighty-four* (1949) in the aftermath of all three events. In Oceania, the Stalinist 'Big Brother' stood for state repression, 'the boot smashing down on the face forever'.

Brave New World proved to be a far more accurate prophecy than *Nineteen Eighty-four*.

Huxley understood how seductive techniques of persuasion could reinforce compliance. In 632 AF ('After Ford'), the five classes in society are taught 'elementary class consciousness', Pavlovian conditioning as they sleep. Today, compliance and persuasion occur while people are awake, largely through the media.

Huxley described a population made docile by brainwashing and dumbing down, so that they 'loved their servitude'. In *Amusing Ourselves to Death* (1986), Neil Postman observed:

> Orwell warns that we will be overcome by an externally imposed oppression. But in Huxley's vision, no Big Brother is required to deprive people of their autonomy, maturity, and history. As he saw it, people will come to love their oppression, to adore the technologies that undo their capacities to think ... What Orwell feared were those who would ban books. What Huxley feared was that there would be no reason to ban a book, for there would be no one who wanted to read one.

Orwell's vision of 1984 is frequently identified with computerisation—but this is to completely misread his book, which was confused on technological matters. Orwell did not appear to have grasped the significance of computers and the future implications of electronic data control.

Orwell's depiction of technology was largely mechanistic and backward-looking, almost Dickensian. However, he scored two hits. One was the use of telescreens as an instrument of surveillance, and the other that government linguists would create a new language to make 'heretical thoughts' impossible. With 'management speak', this has already happened.

The French social philosopher and mystic Simone Weil (1909–1943), who had worked in factories to gain experience, described the deadening life in which 'things play the role of men' in her *Oppression and Liberty* (translated 1972), which I read closely.

I picked up some useful ideas from reading Karl Marx (1818–1883), especially *Grundrisse* (or *Foundations*, to give its English equivalent), written at great speed in 1857–58, intended as the framework of a vast, incomplete work, of which *Das Kapital* was only a part. His writing was passionate and ironic. *Grundrisse* was not published in German until 1941, in Russian until 1969, and in English until 1973. *Grundrisse* included material on alienation, the impact of technology, the economies of time, and his utopian vision.

In *Grundrisse*, Marx was more open, speculative, imaginative than his rigid, dogmatic, authoritarian, and—ultimately— totalitarian followers would admit. But they had, after all, not read it.

Marx was wrong in forecasting many elements of the nineteenth and twentieth century (for example, discounting the role of nationalism), but he seemed eerily perceptive about the twenty-first. Francis Wheen wrote that Marx 'could yet become the most influential thinker of the twenty-first century', and many of his ideas are now accepted by neoconservatives (without acknowledging their source).

As Peter Watson observed in his magisterial *The German Genius*, 'It is not just that what Marx had to say about monopolization, globalization, inequality, and political corruption sounds so pertinent after 150 years, but that we take so much of Marx for granted now, without most of us even knowing it.'

As Watson notes, we accept implicitly that:

- economics is the driving force of human development;
- the social being determines consciousness;
- nations are interdependent; and
- capitalism is a wrecking ball that destroys as it creates, especially in the environment.

In 1882, Marx's son-in-law Paul Lafargue rebuked him for writing something that was not sufficiently 'Marxist'. Marx responded, 'What is certain is that as for me, I am not a Marxist.'

Marx had read Adam Smith's *Wealth of Nations* carefully, and accepted his thesis that there are two basic and fundamentally contradictory forms of employment and time use in society: 'labour/time saving' and 'labour/time absorbing'. Marx dismissed the idea that production and wealth creation could be ends in themselves. His words, in *Grundrisse*, Notebook V, now appear prescient:

> Thus the view [in antiquity], in which the human being appears as the aim of production regardless of his limited national, religious and political character, seems to be very lofty when contrasted to the modern world, in which production appears as the aim of mankind and wealth as the aim of production.

The idea of the rise of a managerial class was picked up and expanded in *Kapital* (III: chapter 23) with his prediction of a white-collared managerial and professional class entirely divorced from the ownership of capital:

> An orchestra conductor need not own the instruments of his orchestra, nor is it within the scope of his duties as conductor to have anything to do with the 'wages' of the other musicians ... The capitalist's work does not originate in the purely capitalist process of production ... [but] from the social form of the labour process.

In *Grundrisse*, Notebook VII, Marx quoted from *The Source and Remedy of the National Difficulties*, a pamphlet published in 1821 by an anonymous English radical, probably Charles Wentworth Dilke (1789–1864), a friend of Keats:

> The first indication of real national wealth and prosperity is that people can work less ... Wealth is liberty—liberty to seek recreation—liberty to enjoy life—liberty to improve the mind: it is *disposable time* [and Marx italicised the words] and nothing more.

This is an elitist view, but it takes a very optimistic view of the human condition. Two centuries later, we have not yet grasped the power of the idea to invest purpose and meaning to life (except as consumers).

The political challenge

Donald Horne argued very perceptively in his important book *The Lucky Country* (1964) that the sheer abundance of our mineral base and 'lucky' elements in our history retarded some aspects of our social, economic, and technological development. People in other countries—the Swedes, Finns, Israelis, Japanese—had to live by their wits (or they would starve if they did not). But Australia always had stuff to dig up and sell, and that determined our concept of value.

I telephoned him to confess that I had plagiarised him out of my unconscious. Characteristically, he responded, 'Funny, but I don't remember the technological bits.'

By the 1970s, many Australian workers were being displaced by technological change, the subject that had preoccupied me for so long. Politicians and trade union leaders were interested in particular cases, but few understood the historic global phenomenon.

I had been a teacher, lawyer, and historian before entering politics, with no formal scientific training. But I was deeply curious, widely read in the history and philosophy of science, and obsessive about finding linkages.

In my first (then called 'maiden') speeches in both the Victorian Legislative Assembly (1972) and the Australian House of Representatives (1978), I discussed the transition to a post-industrial society.[1]

In 1978, Malcolm Fraser's Coalition government set up a Committee of Inquiry into Technological Change in Australia (CITCA), chaired by Sir Rupert Myers, the vice chancellor of the University of New South Wales.

In 1979, I wrote a long submission to the inquiry, 'Implications of a Post Industrial or Post Service Revolution on the Nature of Work', and gave evidence before the committee. The Myers Committee published a four-volume report, *Technological Change in Australia*, in 1980.

I noted that of Australia's 784 members of parliament—state, territorial, and federal—only one had made a submission to CITCA. I then became determined to write a book attempting to predict the impact that the computer/information revolution would have on society.

I began in early 1979, and received valuable advice from the eminent social historian Hugh Stretton, to whom I subsequently dedicated the book. He proposed radical surgery, making it shorter, simplifying the language, changing the order of chapters, and including diagrams. Louise Sweetland was an exemplary editor.

The title was taken from J.S. Bach's Cantata BWV 140, *Wachet auf!* ('Sleepers, wake! A voice is calling, the watchman on the heights is calling …'). It was a challenge, a call to action. By implication, Australians were the sleepers, and that made me, arrogantly, I admit, the watchman.

The cover of *Sleepers, Wake!* featured Henry Moore's sculpture *Atom Piece* (1963–64). This was a model for a larger work, *Nuclear Energy* (1965–66), placed at Chicago University on the site of the world's first 'atomic pile' for sustained and controlled nuclear fission. The image's ambiguity appealed to me. Was it a dome? A skull? A helmet? The triumph of technology? Or a harbinger of something sinister? Moore had invited me to his Hertfordshire estate in Much Hadham in June 1979, where I gave him a rough outline of the book and asked if we could use a photograph of his sculpture, and he agreed at once, waiving any fee.

After modest success for *Sleepers* in the first few months, there was a dramatic change. On 27 October 1982, steel and coal workers from Wollongong/Port Kembla, made redundant by automation, marched on Old Parliament House, Canberra. A door was broken, and workers fought with attendants. The invasion of the parliament symbolised a loss of control, with the media describing it as a serious challenge to Malcolm Fraser's Coalition government.

Suddenly, technological unemployment was big news. In the following week, ABC Television's *Nationwide* (a predecessor of *7.30*) produced a three-part series on unemployment and technology. I was interviewed on each of the programs, and *Sleepers, Wake!* received generous promotion.

In March 1983, less than five months after the march on parliament, Labor won the federal election and Bob Hawke became prime minister. I became a minister, but, significantly, was never promoted to cabinet.

I greatly admired Hawke's skills in analysing a problem, mastering the detail, working out a solution, and then explaining and selling it. However, this admiration was not reciprocated. Indeed, he found me profoundly irritating. The attention generated by *Sleepers, Wake!* was a major factor in this.

Many of my parliamentary colleagues were troubled, irritated, even angered by my use of the term 'post-industrial' to describe Australia's future economy. Surely the employment of thousands of workers in huge factories would be a natural evolution for Australia's workforce? Wouldn't members of industrial trade unions comprise an increasing share of the future labour market?

The answer was 'No', in both cases.

My advocacy of 'sunrise industries', which was extensively reported in the 1983 election campaign, was a second area that irritated Hawke and my colleagues. Sunrise industries were going to be high-technology, brain-based ventures, and included manufacturing plasmas and vaccines, medical and communication technologies (such as the bionic ear), and scientific instruments in astronomy and medicine. They would never be large-scale employers, but could be significant wealth-generators. Inevitably, they would overturn the conventional wisdom about comparative advantages in exports. Wi-Fi was a striking example of a potential sunrise industry, partly based on CSIRO patents, but not developed here.

However, talking about sunrise industries disturbed people still working in smokestack or rustbelt factories.

My prediction of sharply declining employment in manufacturing alarmed manufacturers and trade unions, when Hawke was trying to reassure both sides that jobs would remain secure. If I was correct, he must have been wrong. How could that be?

People found it hard to believe that in the period 1965 to 1982, when 2,060,000 new jobs were created in Australia, not one of them, on a net basis, had been in manufacturing, which lost 150,000 jobs (7.3 per cent) in that period.

When addressing public meetings, I often invited audience members to estimate what proportion of the labour force worked in factories. Most suggested figures between 60 and 70 per cent. When I asked how many had family members working in manufacturing, almost all hands went down.

The high point of employment in manufacturing in Australia was in 1965, at 27.6 per cent. In 1982, when the first edition of *Sleepers* appeared, it was 16.5 per cent, and in 2020, 7.0 per cent (912,500 persons).

Trade union membership comprised 60 per cent of the labour force in 1954, and since then (apart from a short blip in 1961–62) there has been a steady decline, down to 13 per cent in 2019. Now, far more professionals, nurses, teachers, and public servants are trade unionists than blue-collar workers.

In April 1983, prime minister Hawke convened a National Economic Summit Conference (NESC), an unprecedented gathering in the House of Representatives in Canberra, of Commonwealth, state, and local governments, business sectors, and trade unions, churches, and welfare agencies.

As a minister, I had written a carefully argued and documented speech for the NESC, but was specifically dis-invited, out of concern that my argument might frighten the horses.

The NESC had been planned as a stroking exercise, to secure co-operation between opposing forces. It was worthwhile in itself, but none of the great issues that Bob Hawke and Paul Keating are now remembered for appeared in the communiqué published at the end—lowering tariffs, floating the dollar, taxation reform, opening up to a free market, or the global economy. Nor was adapting to the impact of technological change.

What *Sleepers* argued, and the reaction to it

In the early 1980s, Australian economists and the government used a three-part classification of the labour force, in which the Primary sector comprised 'agriculture, forestry and mining', the Secondary was 'manufacturing', and 68 per cent of all workers were, in effect, placed in an undifferentiated residual category called Tertiary 'services'.

This seemed profoundly unhelpful to me, as if religions in Australia could be classified as Muslim, Buddhist, and Others.

I adopted a four-sector classification proposed in the United States by Marc Porat and Edwin Parker (1975) in which Primary was defined as 'extractive' (agriculture, fishing, forestry, mining), Secondary as 'mining and construction', Tertiary as 'services not based on the transfer of information', and Quaternary as 'information' (including printing, which was then, but no longer, a significant employer).

I demonstrated that employment in 'services' had become the largest employment sector in the Australian colonies between the censuses of 1861 and 1871, long before Britain, the United States, Germany, or France, largely due to our exceptional (but barely recognised) degree of urbanisation.

In this four-sector analysis, people working in the collecting, processing, or manipulation of information—teachers, office workers, public servants, managers, and workers in banking, insurance, communications, and media, whose tools of trade were computers or telephones (as they were then called)—were the largest source of employment. They were followed by those providing basic, routine, and repetitive services—cleaning, delivery, retailing, hospitality, tourism (mostly by women, who were often casual and poorly paid).

I then went further, proposing a five-sector analysis of the labour force, with a new Quinary sector covering domestic and quasi-domestic employment (much of it unpaid)—the provision of meals, accommodation, laundry and home cleaning, childcare, repairs and maintenance, gardening, and sexual services. As the two-income family became normative, many women transferred from the domestic economy to the market economy, performing similar services for low or insecure wages.

Ironically, the ALP's television advertisements for the prematurely called 1984 federal election faithfully reproduced employment images of the past—steelworks, farms, construction, cars and railways, heavy engineering—and no emerging ones. There were only a few women in clothing factories, no white-collar workers, and nobody in research or teaching.

Sleepers became both a critical success and a bestseller, going through four editions and 26 reprints in twenty years. It sold 80,000 copies in Australia, and became a textbook. It was praised in the United States by the economist Kenneth Boulding, in Britain by the sociologist Raymond Williams, and in Ireland by the philosopher Charles Handy. It received a glowing review in *The Economist*.

Bill Gates, the founder of Microsoft, had read *Sleepers*, and visited me in Canberra in 1984. He was less than 30 years old. I was impressed by how closely he had examined the text. Bob Hawke and Paul Keating were too busy to fit him in.

Sleepers, Wake! probably had more influence overseas than in Australia. It was translated into Chinese, Japanese, Korean, Swedish—and Braille.

I was interviewed by David Frost and Michael Parkinson for BBC Television, and became the subject of features in *Playboy*, *New*

Scientist, New Statesman, and the *Christian Science Monitor*.

Praise in a special report on Australia in *Nature*, the world's most important scientific journal (in vol. 316, no. 6025, 18–24 July 1985), was no help either.

In 1985, a team of examiners from the OECD (Organisation for Economic Co-operation and Development) visited Australia to examine our strategies for science, technology, and innovation. Its report, *Reviews of National Science and Technology Policy: Australia* (Paris, 1986), endorsed my priorities, but was very critical of a submission from the Department of Finance.

Finance took this very badly, and thereafter I became a target when future budgets were being prepared.

Ministerial frustration

The ALP broke with its long tradition of support for protection when Hawke and Keating courageously reduced tariffs and determined that the Australian economy had to be opened up and able to compete in a global market. Complaints from the motor-manufacturing, textiles, footwear, and clothing industries were met with assurances that, in the long term, Australia's economy would expand significantly—and this happened.

The new direction was zealously encouraged by bureaucrats in Treasury and Finance, and I came to think of it as the 'One Big Idea'.

My argument that research leading to the creation of new industries such as biotechnology and medical instrumentation would require increased support was met with a robust rejection: 'This is exactly the same as the special pleading by manufacturers of shirts and motor vehicles. If we subsidise new industries, we'll have to maintain support for old ones. Leave it all to the market to decide where to allocate resources.'

Unhappily, the Australian market was not well informed about R&D.

Cabinet imposed 'efficiency dividends' through which finite tasks were to be processed more quickly by the use of new technology. This did not make much sense when the CSIRO or the universities were exploring the unknown. Should there be a limitation to the number of constellations observed or parasites detected? Treasury and Finance might well have reduced the Ten Commandments to eight, or directed Michelangelo to choose between painting *The Creation of Man* or *The Last Judgment* in the Sistine Chapel.

Because of the international interest generated by *Sleepers, Wake!*, I became the only Australian minister ever invited to address a G-7 ('Group of Seven': that is, the world's seven-biggest economies) summit meeting, to be held at Meech Lake, near Ottawa, in October 1985.

I wrote to the prime minister's office seeking approval for the visit, assuming this would be automatic and the invitation regarded as further confirmation of the Hawke government's greater glory.

Not a bit of it. My request was smartly rejected. I asked for reconsideration. No, I was told, it was out of the question, I had already been away twice in 1985. Sending me to the G-7 summit would mean exceeding the annual ministerial-visits quota, and this would be politically impossible.

Was there a way out? 'Well', said an official from the prime minister's office, 'you could always take leave and pay your own way to Canada as a private citizen.' And that was what I did. I took leave, and paid for an economy fare to get there. (My accompanying officials travelled first class, because I could authorise their travel but not my own.)

Thirty-four years later, in August 2019, Scott Morrison became the first Australian prime minister ever invited to attend a G-7 summit, held in Biarritz, as a 'special guest'. India, South Africa, and Chile were also invited as guests.

After being invited to address the high-prestige Asia Society in New York and the New England-Australian Business Council in Boston (both in December 1985), I was knocked back once more. So, again, I paid my own way.

South Korea took *Sleepers, Wake!* very seriously, and a government-sponsored translation appeared in 1987. However, limitations on ministerial travel meant that I was refused permission to fly to Seoul for the launch and to address the symposium that followed.

The Irish minister for industry, John Bruton, later the prime minister (taoiseach), used *Sleepers, Wake!* in setting priorities for 'the Celtic Tiger', and spent time with me in Canberra and Dublin.

The city of Sudbury, in Ontario, formerly a mining town, based on nickel, was surrounded by wasteland. Its council sent a delegation to meet me. They planned to transform Sudbury into a science- and culture-based city, creating research facilities, a symphony orchestra, planetarium, science museum, art gallery, and film and music festivals, and *Sleepers, Wake!* would be a users' manual. Could I visit them and advise them? Regrettably, I had to decline.

I was invited to chair the OECD examiners' review of the Yugoslav economy in 1987. After the report's completion, a formal dialogue was planned between panel members and the Yugoslav prime minister, Branco Mikulić, and his senior ministers, in Dubrovnik in October.

Coincidentally, Bob Hawke, touring Europe with a gaggle of Australian journalists, happened to be in Dubrovnik on the same day for a scheduled meeting with Mikulić.

Hawkie was told that his discussion with his Yugoslav counterpart would have to be truncated.

'Why?' he asked.

'Mikulić has to meet with Barry Jones.'

'Barry fucking Jones? Here in Dubrovnik?'

When the OECD commissioned me to carry out assignments in Venice (1985), Paris (1987), and Montréal (1991), in each case *Sleepers* was cited as the justification.

Sleepers was translated into Chinese by three academics at the University of Western China, in Chengdu, Sichuan, and published in 1988.

On separate occasions, Deng Xiaoping's two daughters visited me in Canberra—Deng Rong in 1988, and Deng Nan in 1994—to tell me that their father had read *Sleepers, Wake!* and it had influenced his thinking. I found this hard to believe, but still …

In 1990, I was invited to be part of an international think tank in Moscow to advise Mikhail Gorbachev on Perestroika. Could the Commonwealth government support me on this project? The answer was 'No'. However, my friend Emanuel Klein was able to get funding from Sir Peter Abeles, an intimate of Hawke. (I doubt if he ever knew.)

Labour/time-use value

Sleepers identified the problem of redefining 'labour/time-use value', and putting it on the political agenda, as well as into education practice.

Time budgeting and self-management of time are central to personal development, from infancy on. Time is the medium in which we live: the only irreplaceable resource. Using it effectively involves setting priorities. But there is a paradox: time management, historically, has been an instrument of external control. We find it extraordinarily difficult to impute a value to our own time.

I was careful not to predict the collapse of work, which many writers on future trends had glumly asserted.

Three economic and time-use eras

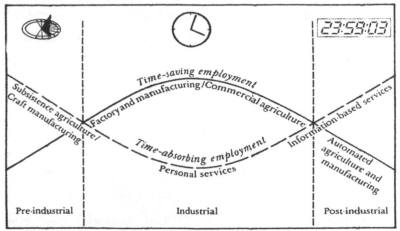

Source: Sleepers, Wake!

I thought that service employment had an almost infinite capacity for expansion. For example, Australia has far more hairdressers than steelworkers. I acknowledged that many of the new jobs would be part-time, low-paid, and often unsatisfying, in what has been described as 'the bed-pan economy'.

The Australian colonies began adopting the eight-hour day (8 x 6 = 48 hours) in the 1850s. The 44-hour week became law in 1927, and the 40-hour week in 1947. Economists predicted that by the year 2000 the working week would have fallen to 28 hours.

In 1983, the standard Australian working week was reduced to 38 hours, but since then, taking overtime into account, hours of work for full-time employees have plateaued.

In the pre-coronavirus world, Australia had among the world's longest working weeks, 42.3 hours (including overtime) for full-time males. In addition, commuting periods were unusually long, and most Australians were engaged in employment-based activity far longer than their grandparents were. More than a million Australians worked for six, or even seven, days each week.

Job creation, in normal circumstances, is easy—just imagine

how many jobs could be created if we abolished reticulated water supply and sewage-disposal systems, and replaced tractors and earth movers with an army of spade-wielding diggers—but it would be stupid and pointless to do so.

The problem is income creation.

Post-industrial society

I argued along parallel lines to the American social scientist Peter Drucker that the post-industrial society would be the first in which the great majority of the labour force would no longer be performing the same work in factories, offices, or fields. This would be far more than a social change. It would be a change in the human condition.

In a knowledge society, having a hundred or a thousand people performing the same task would be pointless. A hundred or a thousand people can and must create a hundred or a thousand different types of production, all highly individualised. People will make their own work, depending less on being a servant to a master. Knowledge workers, Drucker noted, are already the leading (not ruling) class of the knowledge society. This poses particular problems for the unskilled unemployed.

In the fourth edition of *Sleepers, Wake!* I made a brave attempt (on pp. 38–41) to identify the 'X' industry that would arise after the initial impact of the post-industrial revolution, and provide future job opportunities:

> [N]ew forms of entertainment; the expansion of sport; tourism and the hospitality industry; combating disease; maintenance services for extreme longevity; personalized care of the handicapped; raising literacy standards; developing new forms of waste disposal; improving water quality; restoring soils; tree planting; pollution control; developing high-value crops as

economic alternatives to devastating rain forests for short-term economic advantage; adapting rice farming and cattle digestion to produce less methane; and designing new fuels and industrial materials with reduced environmental impact. [I have reworded this from the original, which read awkwardly.]

The list stands up pretty well.

With a rapidly growing and ageing population, there have been increased numbers of physicians, dentists, and specialists such as oncologists, cardiologists, and psychiatrists; nurses, midwives, and hospital administrators; and medical researchers, in which field Australia has a very high international ranking. Health would also include programs for fitness (even 'wellness', if that word can be excused) and rehabilitation.

There has been a significant growth in work for 'tradies': plumbers, electricians, and locksmiths, and others such as those who fit solar panels, and maintain or repair air conditioners and heating services. In a previous generation, these skills would have been employed in factory work. But now 'tradies' are independent contractors, non-unionised, and, it would appear, often Coalition voters.

Back in the 1980s, I should also have listed accounting, financial services, and wealth management as areas for future employment growth, and put more emphasis on tourism. However, in my defence, I did identify significant tourist-related activities, such as gambling, accommodation, sporting events, access to natural wonders, eating out, and the provision of drugs, drink, and sex.

Back then, I argued that Australia suffered from an 'inventory' problem, the lack of high-value-added brand-name goods for which there was an international demand. This is still the case. We measure our exports in tonnes, but import high-priced goods weighed in grams. The steel in a Seiko watch would have been fabricated from Australian iron ore. But what was the value of the Australian contribution? 0.01 cents—or less?

I remember one occasion when I went to Lismore, New South Wales, to deliver a speech. I flew up in a Saab, was picked up in a Volvo, was photographed by a Hasselblad camera, used an Ericsson phone, and drank from an Orrefors glass—all Swedish products. It was hard to imagine flying to a Swedish town in a Nomad, being picked up in a Holden, being photographed by a box brownie, and drinking from a Vegemite jar.

In Chapter 9, 'Work in an Age of Automata', I discussed the ideology of work, the concept of 'work for work's sake', a punitive approach to labour that is a central mantra not only of capitalism, but of most religions, too.

In the Judaeo-Christian tradition, work was imposed as Adam's curse. In Greek mythology, work was an element that escaped from Pandora's box, and work was πόνος (ponos), the root from which the words 'pain' and 'punishment' are derived.

This is central to Scott Morrison's repetitive, 'If you have a go, you'll get a go!' and, as John Lanchester put it, 'our culture's deeply imbued ideas about the innate and redemptive virtue of paid work'. It is the ethos of Newstart, a social program with some punitive features.

The story of Jesus, Martha, and Mary (Luke x, 38–42) lends no support to the work ethic. However, the parable of the talents (Matthew xxv, 14–39) is in the authentic spirit of capitalism, and Paul's words, 'For even when we were with you, this we commanded you, that if any would not work, neither should he eat' (II Thessalonians iii, 10) go even further.

After seven years, I was defenestrated from the ministry in the wake of the 1990 election. There was a need for fresh blood, I was told. Well, someone's blood, anyway. As it turned out, I was replaced by two-and-a-half ministers (Simon Crean, David Beddall, and

Peter Staples) who handled my responsibilities for science, customs, small business, and—oddly—housing policy.

There were several reasons for my relegation: Hawke's lack of support; the emergence of nation-wide factions that squeezed out non-aligned ministers; and the fact that the ALP had lost six seats in Victoria, a state that was regarded as over-represented in the ministry.

However, I had some grim satisfaction that I had prevented the CSIRO from being broken up into sectors with a direct fee-for-service relationship with client groups in agriculture, manufacturing, and IT; secured funding for Australian participation in radio-astronomy; and established the Australia Prize for Science, which still survives as 'the Prime Minister's Science Prize'. (I regretted the name change: there are hundreds of prime ministers, but only one Australia.)

I treasured my experiences in Antarctica, and was gratified when Barry Jones Bay (69° 25' 30' S, 76° 03' 00' E) in the Larsemann Hills area was named for me in 1988.

I liked to think that my nagging Hawke about Antarctica contributed to his late, passionate, and successful campaign—against the odds—to secure international agreement (the Madrid Protocol, in 1991) for a 50-year moratorium on exploration for minerals and oil.

Long-term strategies

I remained in the parliament, and received two consolation prizes: to chair the Australian House of Representatives Committee for Long Term Strategies (1990–96) and to represent Australia at UNESCO and the World Heritage Committee in Paris (1991–96), the latter essentially as a commuter.

The long-term strategies committee, with bipartisan membership, produced four major reports, all arising from major elements in *Sleepers, Wake!* All were adopted unanimously.

They were 'Australia as an Information Society: grasping new paradigms' (1991); 'The Role of Libraries/Information Networks' (1991); 'Expectations of Life: increasing the options for the 21st century' (1992); and 'Australia's Population "Carrying Capacity": one nation—two ecologies' (1994).

Longevity

Longevity was a significant element in *Sleepers, Wake!*

The English historian and social scientist Peter Laslett had written perceptively about the emergence of a new demographic group, 'the Third Age', who lived healthily for decades after retirement.

The University of the Third Age ('Université du troisième âge') had been created by Pierre Vellas in Toulouse in 1973, and in 1982 Peter adapted the concept and was co-founder of the first British version. While I was preparing our committee's third report, on expectations of life, Peter became a valued consultant and, later, friend.

In 1999, I was elected as a Visiting Fellow Commoner at Trinity College, Cambridge, and worked with Peter in 2000 and 2001 on political and social problems associated with the Third Age. After retiring from parliament in 1998, I took up the cause of the Third Age with enthusiasm, and worked with Peter at the Cambridge Group for the History of Population and Social Structure in 2000 and 2001.

I am now five years older than Queen Victoria was when she died, and I hope I don't look it. But I can remember a time when a typical male worked until compulsory retirement at 65, often with employment-related disabilities, and then died around 70, having spent his entire life in the same locality he had been born in.

By 2015 I had become the longest-living male in my family's history, going back to Adam (or Eve), just as I had been the first to graduate, and—war service aside—the first to experience life outside my own country. This experience was commonplace among my contemporaries.

With following generations, the prospects for diversity of experience, overseas travel, and a very long life were taken as axiomatic. Most Australians can now anticipate a period in retirement almost as long as their working lives—an unprecedented phenomenon in population history, with significant social, economic, and political implications. People will be living longer, probably working more hours each week (including commuting time), but for a reduced percentage of total life.

If we assume a life expectancy of 83 years, then 40 years of paid work, for a 42-hour week, 48 weeks a year, accounts for 11.09 per cent of a total lifetime.

The political priority for a growing proportion of older citizens is now preservation of the value of family assets.

So what?

In the fourth edition of *Sleepers,* the most extensively revised material dealt with education, especially in the context of lifelong learning.

I canvassed whether societies could be both tolerant and competitive, and whether schools could cope with an exponentially increased knowledge base, even with computers.

Somebody wittily described the reaction to *Sleepers, Wake!* as having been 'What?' in 1982 and 'So what?' in later decades. I scored well in my judgements of how technology would transform the world. However, the rate of take-up was much faster than I had anticipated, and, like most other writers, I failed to anticipate the impact of negative aspects of the Information Revolution.

In 1982, I gave a speech in Hobart predicting that by the year 2000 there would be more computers in Tasmania than motor vehicles. The Hobart *Mercury* regarded this as so ludicrous that it editorialised I should be put on lighter duties. Years later, I met a man who had cut out the editorial and kept it in his wallet, determining that if we ever met and I was correct, he would give me a crayfish.

We did, I was, and the crayfish was delicious.

The information divide: the complexity problem

In the 1970s, I had begun arguing that the deepest division in society would be between 'the information rich' and 'the information poor.' I assumed that access to new technology would open people up to the world, that they would explore the universal and long term, although I did quote T.S. Eliot, in his choruses from 'The Rock' (1934):

> Where is the Life we have lost in living?
> Where is the wisdom we have lost in knowledge?
> Where is the knowledge we have lost in information?

Access to data has increased exponentially, at a sharply reducing cost, with gigabytes being transmitted for cents. Now, society's deepest division is between 'the data rich' and 'the knowledge poor', with access to wisdom and reflection being blindsided.

There are inbuilt tensions between the nature of major challenges and our failure to understand or address them:

- Our political cycles are short term (Australia, for example, operates with a three-year parliament for the Commonwealth, and three- or four-year parliaments for the states);

- Media cycles are very short term (news editors get very tired of a story after 24 hours or so);
- Digital media cycles are even shorter (turnaround time is measured in hours and minutes); and
- The attention span of social media is shorter still (with messages often having a half-life in seconds). Understanding complex issues is outside the range of social media, where the emphasis, notably with Twitter, is on immediate reactions.

The great Harvard biologist E.O. Wilson observed that humans have 'palaeolithic emotions, medieval institutions, and God-like technologies'.

Two editions of *Sleepers* appeared while I was a minister. This inhibited my criticism of government failures in reacting to scientific and technological challenges.

I annotated the material entitled 'a political program for survival and enhanced quality of life' to indicate that of the 64 policy recommendations I had made in 1982, only eighteen had been adopted by 1995. It was a matter of bitter frustration to me that there was, and still is, such strong resistance to making science a government priority. We responded slowly, and passively, to the Information Revolution and the rise of biotechnology, and failed to adopt imaginative strategies that could have strengthened our economy, and promoted innovative education, social welfare, and trade policies. *Sleepers* had a major impact, but a slow one, as graduates who had studied it at school and university adjusted to an intellectual environment dominated by technology—little of it originating in Australia.

Of course, many of my policy recommendations are now obsolete. Those intended to protect the then existing labour force and to placate trade unions were cumbersome, bureaucratic, and rigid. Legislating for full employment (even as a social goal) would be as pointless as attempting to legislate for longevity or to regulate tidal surges.

My recommendation for a National Information Policy was ignored: a very serious failure by government. Adopting such a policy would have required Commonwealth, state, and local governments, and the corporate sector, public and private institutions, and citizens generally to think seriously about the implications, opportunities, challenges, and dangers of the digital revolution.

But we did not, and are now paying a price for that.

A National Information Policy had been recommended, without dissent, by the all-party House of Representatives long-term strategies committee in 1991.

In 2000, Kim Beazley, then opposition leader, nominated me to chair the Knowledge Nation Taskforce. Martyn Evans, then the shadow minister for science and resources, was appointed as deputy chair. Other members, all chosen by the Beazley office, included John Brumby, Craig Emerson, Carmen Lawrence, Evan Thornley, Tracey Ellery, and Steven Smith.[2] I proposed one more name, Kevin Rudd, and this was agreed to.

Dr Dennis Perry was our resourceful technical advisor. Two members (Ellery and Thornley) flew from California for meetings, but some, including my deputy, were infrequent attenders and inactive contributors.

We invited public submissions, and received significant input from Peter Doherty, Peter Karmel, Fiona Stanley, Gus Nossal, and others.

Our report, *An Agenda for the Knowledge Nation* (2001), agreed to unanimously, was a comprehensive policy framework linking those elements in Australia's society, economy, and environment, especially human and physical resources, dependent on the generation, use, and exchange of knowledge.

Running to 58 printed pages, it proposed policy recommendations for the 2001 election, included a National Information Policy, and followed up some of my ideas from *Sleepers, Wake!* about research, education, and the creation of new industries.

It did not propose increased government activity, but argued that the government should act creatively as a catalyst, encourager, information and infrastructure provider, major customer, and exemplar of world's best practice. The government could also define and assert the concept of 'the public good'—that education and health, for example, are not just commercial enterprises, determined by market forces. We proposed that the government become a change agent in encouraging private-sector investment in new knowledge-intensive industries.

Education was a central concern, but the report emphasised the importance of linkages and the nature of complex systems. We concluded that much important knowledge was locked up in silos and that national connections were very weak, compounded by the dispersal of population across a huge continent and rivalry between states.

The 'complexity diagram', the most mocked and least understood part of Knowledge Nation, illustrated the most important single element in the document: Knowledge Nation (KN) was a central unifying principle—not an institution—with complex, dynamic interactions.

The government used the techniques of 'spin' to ridicule the Taskforce Report, and they succeeded brilliantly. The attack, not the report's contents, then became the story, compounded by Labor's failure to defend Knowledge Nation, let alone argue for it. The main attack was on my notorious 'complexity diagram', in my introduction, not in the report itself. The diagram was a 'mind map', a schema devised by Tony Buzan in England in 1968, and widely used in schools as a graphic technique to illustrate complex relationships.

We proposed a Knowledge Bank, providing access to all

databases held by the Commonwealth, states, and territories, for which I revived the old word 'cadastre' as the basis for evidence-based policy formulation, and this became the second target for attack. (I concede that the World Wide Web makes it easy to access vast amounts of material, but most material is locked away in silos, and a 'cadastre' could have been valuable in linking, say, Australia's forest cover and water supply.)

The diagram was repeatedly attacked, but nobody, to my knowledge, attacked the proposition that it illustrated. The proposition was never discussed by our critics or even by the party. Young people, familiar with mind maps, understood my diagram; journalists and politicians did not. To many columnists and cartoonists, the diagram was the beginning and the end of discussion.

Oddly, the diagram has a family resemblance to Google's illustration of its PageRank 'ranking algorithm', the operating principle of the world's dominant search engine.

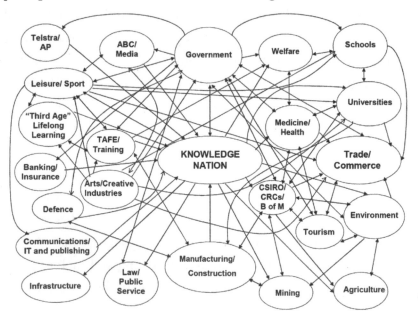

Source: An Agenda for the Knowledge Nation

Peter Costello, the treasurer, began the attack, dismissing our report as 'Noodle Nation', and his description of the diagram as 'spaghetti and meatballs' had some wit. Brian Loughnane, later the Liberal Party's federal director, coined 'Noodle Nation', while 'spaghetti and meatballs' came from Costello's media advisor, Niki Savva (who later graciously expressed regret for her part in the exercise). All three scored a direct hit, and David Kemp, the education minister, pursued the attack. This clever tactic, a classic example of spin, meant that the issues raised in the report were never debated.

The Australian obviously thought the diagram was hilarious—and published six cartoons by Peter Nicholson based on it. The newspaper's Matt Price described the complexity diagram as 'crazy, incomprehensible and infamous'. I thought that 'infamous' was rather strong. Both responses reflected a wilful refusal to even debate the report: *Don't debate the contents, just destroy it.*

Kim Beazley's office failed to respond. Their line was, in effect, *If we say nothing, the issue will go away.* It didn't.

Ultimately, the contents of our report were scuttled by the leader's office before polling day. However, bizarrely, the phrase 'Kim Beazley's Knowledge Nation' was used as the centrepiece of the ALP's policy launch. It confined itself to proposing modest increases in education funding over five years. Weirdly, in a table headed 'Value of Labor's Commitment', projected expenditure for the Knowledge Bank (or cadastre) was listed as '$0.0—spread over five years'.

Crikey.com (in October 2004) contained a story 'from a well-placed Labor heavy' that described Knowledge Nation as being all my own work, and as 'a policy so obscure and obtuse' that insiders knew it would be a 'laughing stock'. The only example given were the notorious complexity diagram, which was, inevitably, complex, and revived the word 'cadastre' to describe the Knowledge Bank.

The report finished with an anecdote expressing regret by Beazley staffers that they had failed to run me down in a car in a darkened Melbourne street 'because no jury would have convicted them when the real story came out'.

Climate change/global warming and political paralysis

Sleepers, Wake! was a major achievement. It was written so long ago that I can read it objectively, because the author seems to be somebody whom I once knew but lost touch with.

Coming across unfamiliar paragraphs, I cannot help thinking, *The author was very perceptive. I wonder what happened to him?* However, *Sleepers, Wake!* contained a major omission. I did not address the issue of climate change/global warming. I was well aware of its importance, but decided to concentrate on the transition from industrial to post-industrial employment, science as a transformational force, the information revolution, time use, and education. That seemed to be enough for one book.

I had first become concerned about the risks of climate change/ global warming in 1967, more than half a century ago, when I interviewed the distinguished Scottish science writer (Lord) Ritchie Calder on the subject on a talkback program on commercial radio in Melbourne.

In the United States, James Hansen of the National Aeronautics and Space Administration (NASA) drew President Jimmy Carter's attention to climate change (without much success) in 1979. George H.W. Bush promised to tackle climate change in his campaign for the presidency in 1988, but was then talked out of it. In the late 1980s, Al Gore, then a Democratic Party US senator from Tennessee, took up the issue seriously.

I began talking and writing about the challenge of climate change/global warming in 1984, and was one of the first Australian politicians to address the issue. This claim is less audacious than it sounds. For some years I was the only competitor in the race, and a fat lot of good it did me, politically.

In 1985, cabinet agreed to set up the Commission for the Future (CFF), which proved to be highly controversial, and my old mate Phillip Adams, columnist, broadcaster, and film producer, became its foundation chair. The commission was not an exercise in futurology—its aim was to provide information and to generate serious community debate about future policy options.

The CFF commissioned and published some valuable papers. The *Personal Action Guide for the Earth* (1989), a powerful analysis of greenhouse material, was awarded a Gold Medal by the United Nations, and was widely quoted internationally.

'Greenhouse '88' was a series of conferences in all of Australia's capital cities and in Cairns, jointly organised by the CSIRO and the CFF. However, the CFF was weakened after I left the government, and was ultimately given to Monash University. (It was put to sleep in 1998.) As it turned out, the Australian Senate took a stronger position on the greenhouse challenge in 1989 than it did in 2019.

I was struck by the differing responses to two serious climate challenges: the human contribution to global warming due to greenhouse gases, and the impact of chlorofluorocarbons (CFCs) in creating a hole in the ozone layer, increasing exposure to ultra-violet light from the Sun, and a higher incidence of cancer, especially melanoma.

The first proved almost impossible; the second was easy.

Britain's prime minister, Margaret Thatcher, convened a Saving the Ozone Layer conference, to be held in London in March 1989. She invited Senator Gore and me to be the keynote speakers. *More recognition for Australia*, I thought. *The prime minister is sure to say 'Yes'*. As required by the rules, as a minister I sought his approval to attend and, once again, Bob Hawke said 'No'.

When I told Sir John Coles, the British high commissioner to Canberra, that I would have to decline Mrs Thatcher's invitation, his face darkened. He said, 'She particularly wants you to speak, and if she is thwarted, well, … it can be ugly. Would it be acceptable to your government if the UK picked up your air fare and expenses?' Bob Hawke's office then agreed, rather grumpily, that I could attend.

The international community, and the major chemical corporations, readily accepted the argument that CFCs used as propellants in aerosol sprays were depleting the ozone layer, although their volume as a percentage of the atmosphere is minuscule compared to carbon dioxide and methane. This is in striking contrast to the combination of fury, hysteria, and mendacity that has greeted the evidence of climate change/global warming.

The explanation was that, in the case of CFCs, every chemical corporation grasped that there were economic advantages in getting into the market first with HFCs (hydrofluorocarbons) as an alternative propellant. However, to most of the fossil-fuel industry, the global-warming challenge is a fight to the death.

Climate change is the great moral challenge for this generation, with profound implications about how we make choices. Short term or long term? Consumption or preservation? Ourselves or our descendants? Rejecting or examining evidence and expertise? Emotion or rationality in making political choices?

Climate change is inextricably linked to the probable irreversible extinction of species, and the enlargement of the world's arid zones—the central factor in the displacement of millions of climate refugees, which has led to a paralysis in our politics, a loss of faith in our institutions and the Western liberal tradition, and the rise of authoritarian governments, snarling nationalism, and, often, irrational leaders.

These issues must now be addressed.

CHAPTER TWO

Democracy's Existential Crisis in a Post-truth Era

Things fall apart; the centre cannot hold ...
The best lack all conviction, while the worst
Are full of passionate intensity.
– W.B. YEATS, 'THE SECOND COMING' (1919)

Historians and political scientists have classified major events in recent world history into two periods, 1901 to 1945 (44 years), with the end of World War II as the dividing line, and 1945 to 2020 (75 years).

The period 1901 to 1945 was marked by aggressive nationalism; trade wars; high tariffs; brutal colonisation; World War I; the rise of totalitarian rule in Russia, Italy, and Germany; the Great Depression; World War II; the Holocaust; and the atomic bomb.

From 1945 to the present, 'the post-World War II structure of interlocking diplomatic, military, and economic agreements and organisations ... have preserved peace, stability, and prosperity', as Christopher Browning put it in the *New York Review of Books* (25 October 2018). People are living far longer, even in the Third World (life expectancy globally is now 70.5 years), and despite rapid population growth, infant mortality has fallen. Female liberation still has a long way to go, especially in India, but is much improved, and the threat of global war is remote. The British Empire ceased to exist. China, India, Pakistan, Indonesia, and South Africa underwent enormous changes.

However, the 1945–2020 era has had its downside: wars in Korea, Vietnam, Afghanistan, and Iraq; civil wars in parts of Africa, the Middle East, Asia, and the Balkans; the Cold War, with nuclear threats and Stalinist control of Eastern Europe until 1989; Mao's purges and famines in China; terrorism; and the mass displacement of refugees. Local warlordism remains intact in parts of Africa, Asia, and the Middle East. Corrupt regimes are commonplace. Slavery survives, but is in terminal decline. Increased consumption levels are destroying the environment and polluting the air, sea, and land.

But there has been no World War III, and the atomic bomb was never used after Hiroshima and Nagasaki in 1945.

The post-World War II arrangements are now under increasing threat, largely from Donald Trump and his wrecking-ball approach to the United Nations, the EU (European Union), NATO (North Atlantic Treaty Organization), the G-8 ('The Group of 8'), the G-20 ('The Group of 20'), the WTO (World Trade Organization), the World Health Organization (WHO), the IPCC (Intergovernmental Panel on Climate Change), the IMF (International Monetary Fund), and any other institutions that attempt to address global issues. Vladimir Putin has also been an important destructive agent.

We can see the rise not of totalitarianism, but of 'illiberal democracy', a model that operates in Russia, Turkey, Hungary, Poland, and Brazil. China is a special case. Vested interest is preferenced above public interest, illustrated by the refusal to tackle the problem of climate change seriously.

The politics of anger and resentment has displaced the politics of rationality and optimism.

In 1950, as a first-year student at Melbourne University, I observed, at close quarters, the great English philosopher Bertrand

Russell (1872–1970), like H.G. Wells no longer a household name. He said, 'Three passions, simple, but overwhelmingly strong, have governed my life: the longing for love, the search for knowledge, and unbearable pity for the suffering of mankind.' I have tried to live by his words.

The Polish political philosopher Leszek Kołakowski (1927–2009), exiled in Oxford from 1968, wrote in 1982: 'Social democracy requires in addition to commitment to a number of basic values, hard knowledge and rational calculation … it is an obstinate will to erode by inches the conditions that produce avoidable suffering, oppression, wars, racial and national hatred, insatiable greed and vindictive envy.'

This formulation resonates with me.

Democracy: history, practice, and ethos

Liberal democracy currently faces its greatest existential crisis since the 1930s: it is being displaced by emotive and often poorly informed populism internally, and a strident nationalism externally.

What is sometimes called 'the Enlightenment project' has come under sustained attack in the United States, much of Europe and, to a lesser degree, so far, Australia. Democracy is a very fragile flower.

The adoption of universal suffrage, without any discrimination on the basis of gender, race, or property qualifications, has been comparatively recent—after World Wars I and II, and the end of imperialism.

In the eighteenth century, there were early experiments with female suffrage in Sweden-Finland, Corsica, Lower Canada, and New Jersey.

In the nineteenth century, Wyoming (1869), Colorado (1893), and Idaho (1896) in the United States, New Zealand (1893), South Australia (1894), and Western Australia (1899) were pioneers in granting women the right to vote.

Raw frontier societies that put heavy demands on settlers to create new structures were supportive of female suffrage. Britain's far-flung colonies were far ahead of the mother country in experimenting with democracy.

The provision of compulsory primary education in Great Britain, much of Europe, the US, Canada, Australia, and New Zealand in the 1870s was regarded as a necessary step towards personal development and democratic practice. However, while Germany may have been the best-educated nation in Europe, the democratic ethos failed to flourish there. That is because there is a profound difference between democratic practice and democratic *ethos*.

In 1914, more males, proportionally, were entitled to vote in Imperial Germany and Imperial Austria than in Great Britain, although Britain had more of a liberal ethos than either. So Britain's rallying cry at the outbreak of World War I was not for 'Democracy' but for 'King and Country'.

In April 1917, President Woodrow Wilson asked Congress to bring the United States into the war. He used the words 'The world must be made safe for democracy' for the first time.

Britain's general election of 1918 was the first with universal male suffrage, and women over the age of 30 were given the right to vote.

Universal suffrage was very rare until after the end of World War I. The sequence in which nations adopted universal suffrage deserves serious examination and reflection, especially the gap between rhetoric and practice.

1893: New Zealand. However, women were excluded as candidates until 1919. Māori males could vote from 1867.
1902: Australia. Women could now vote and stand for election, but none was elected until 1943. Australian Aborigines

and Torres Strait Islanders were generally excluded from voting until 1967 (exactly a century after Māori males were enfranchised). Manhood suffrage for white males had been granted in the colonies in the 1850s.

1906: Finland

1913: Norway

1915: Denmark

1917: Russia. Universal suffrage was adopted by the Provisional Government, led by Aleksandr Kerensky, after the tsar's abdication, and before Lenin's seizure of power and the USSR (Union of Soviet Socialist Republics) was established. However, the democratic ethos was missing. No election since then has been free and fair, with the possible exception of the close-run 1996 presidential contest between Boris Yeltsin and Gennady Zyuganov.

1918: Canada. However, Inuit and 'status Indians' (that is, First Nations people who claimed territorial rights) were excluded from voting until 1960. Québec gave women the vote in 1940.

1918: Austria. However, this became inoperative under Hitler from 1938 to 1945. Universal male suffrage had been established in 1896, and some women allowed to vote in 1907.

1919: Germany, although inoperative under Hitler between 1933 and 1945. Universal male suffrage had been granted very early, in 1871, when Bismarck became chancellor of a united Germany.

1920: The United States. Females were given the vote, but Afro-Americans were excluded in many states until 1965. Each state elects two senators, giving disproportionate weight to states with small populations. California, with 40,000,000 people, elects two US senators: the 21 least populous states, which between them have fewer people than California, elect 42 senators. The Electoral College chose George W. Bush and Donald Trump as president, despite both candidates having lost the popular vote. In many states, there are strong measures to discourage African-Americans and former prisoners from voting. Holding elections on a working day is another factor, and turnout is often low, especially in mid-term elections. Turnout was very low in 2014 at 36.7 per cent, but comparatively high in 2018 with 50.3 per cent.

1921: Sweden. Some feudal elements survived in Sweden until 1885, and votes were weighted to reflect wealth (up to a ratio of 54:1). It was the last Scandinavian country to adopt universal suffrage.

1923: Ireland

1928: United Kingdom. Universal male suffrage was legislated in 1918 and women over 30 were allowed to vote. In 1928 all women aged above 21 were enfranchised. The Reform Bill of 1832 was the first serious attempt to rationalise representation on the House of Commons. The House of Lords still retains the power to delay – but not defeat – legislation.

1931 Brazil. The voting age was reduced to 18 years, and women were enfranchised. Democracy was suspended in 1964 and restored in 1985.

1933: Spain

1934: Turkey. One of Kemal Atatürk's reforms.

1944: France. Vote given to males aged 25 in 1792, and then the franchise was restricted until restored to all males in 1848. It was almost a century before women received the vote.

1945: Italy. Universal male suffrage in 1912.

1945: Japan

1947: India. Universal suffrage was implemented from 1950. India is the most populous nation using democratic forms. However, a strident populist nationalism has led to harsh treatment of Muslim minorities.

1955: Indonesia

1971: Switzerland. Women were granted the vote federally by referendum: locally, one canton (Appenzell Innerrhoden) resisted until 1990.

In 2018, the Democracy Index, compiled by the Economist Intelligence Unit (EIU), examined 167 sovereign states (admittedly from a Western liberal perspective), and identified only twenty 'full democracies', including the United Kingdom, Canada, Australia, New Zealand, Ireland, Germany, the Scandinavian states plus Finland and Iceland, the Netherlands, and Spain. The United States, Japan, France, Italy, Greece, Israel, Argentina, and South Korea are classified as a 'flawed democracies'. There were 38 'hybrid regimes', including India, Indonesia, Brazil, Nigeria, Turkey, and Papua-New Guinea, and 52 'authoritarian' ones, including China, Russia, Saudi Arabia, Iran, Vietnam, the Democratic Republic of Congo, and North Korea.

Around the world, there has been a rise of illiberal democracies and authoritarian or kleptocratic rule, corrupted elections, restrictions on free speech, infantilisation of debate, suspension of the rule of law, a resort to violence, and adoption of the surveillance state.

The commitment to Western liberal values has been weakened or even abandoned. Instead of governing in the public interest, narcissistic leaders shout mantras.

Populism

Despite high levels of formal education, there are significant populist elements in contemporary politics. 'Populism' asserts and exploits conflict between the experience, expectations, and desires of the population generally (sometimes condescendingly described as 'ordinary people') and elites, especially professionals with expert knowledge.

Although the masses-versus-elites dichotomy sounds like the historic argument from the left demanding that power move *downwards* in the social pyramid, in contemporary politics 'populism' is passionately promoted by supporters of the status quo, including the gambling, junk-food, and fossil-fuel lobbies, and most of the print media, commercial radio, and television, asserting that elitists/experts exercise too much power.

The essence of populism involves:

- telling people what they want to hear;
- appealing to identity ('I am a white English-speaking male and I don't relate much to other groups.');
- promoting patriotism as the primary value;
- exploiting fears, phobias, and racial memories;
- insisting that one's own people are the best (depending on the audience);
- persuading citizens that they are surrounded by enemies;
- proposing simple solutions to complex problems;
- treating evidence as irrelevant;
- rejecting expertise;
- rejecting alternative points of view;
- promoting tribalism, as with sport; and
- pursuing very selective readings of history.

Challenges to liberal democracy

Liberal democratic practice was profoundly challenged by dramatic events in 1979, 1989, 2001, 2003, and 2016 that transformed politics and economics throughout the world. The rule of law, the commitment to open government, and the rational analysis of policy options became subverted. What democratically elected governments did to themselves went much further than their collective enemies could have anticipated.

I can identify sixteen inter-related factors that have fundamentally challenged the relevance of truth and evidence in how society functions. There may be more.

Islamic fundamentalism (1979–)

From the 1950s, figures such as Sayyid Qutb (1906–66), an Egyptian scholar and theorist, were central in reviving a puritanical and fundamentalist Islam, including the cult of martyrdom. Qutb had become repulsed by Western values, especially the Enlightenment, while on a scholarship in Colorado (1948–50). He denounced the Egyptian ruler Gamal Nasser as a secularist, and was martyred himself.

His prophetic role and zeal for 'jihad' (struggle) against the enemies of Islam parallels the teachings of the Wahabis in Saudi Arabia and the Ayatollah Ruhollah Khomeini in Iran.

In 1979, the overthrow of the Shah of Iran and the Pahlevi dynasty, and Khomeini's proclamation of an Islamic state, marked the revival of Islamic fundamentalism as a political and ideological force, and the concept of 'jihadism' spread dramatically.

This came as a complete shock to Western governments and their intelligence agencies. As Christopher Andrew wrote in *The Secret World* (2018, on p. 700):

Western intelligence agencies at the end of the Cold War suffered, although they did not realise it, from a serious lack of theologians. During the Second World War and the Cold War, they had been well versed in Nazi and Communist ideology. But the increasingly secularized late-twentieth-century West found it far more difficult to grasp the appeal of Islamic fundamentalism.

Islamic fundamentalism was a powerful reaction not only to the political and economic power of the West, but even more to its cultural values, including secularism, materialism, overt sexuality, and equal opportunity for women.

At its extremes, it led to violent attacks on the West, notably on 11 September 2001, in New York and Washington, but also in Bali, London, Manchester, Paris, Nairobi, Madrid, and Mumbai, and in Russia and Syria.

Leaders in the West needed access not just to theologians but historians, psychologists, and linguists as well.[1]

The US and Britain had helped to create the Mujahideen in Afghanistan after 1979 as an instrument to incite Afghans to rise and force the withdrawal of Russian occupying forces. This led directly to the rise of the Taliban from 1991.

In Jordan, the West funded and encouraged Hezbollah (jihadist and clean) to weaken Hamas (pragmatic and corrupt) in Jordan, presumably hoping that a fractured leadership would be helpful for Israel. Jordan had been essentially secular, but jihadism became a significant force.

The Taliban, and later the Islamic State (a.k.a. IS or ISIS) committed acts of barbaric cruelty, including beheadings and crucifixions, and hundreds of civilians were killed by suicide bombers.

Suicide bombers were incomprehensible in the West. Routinely categorising them as 'cowards' showed a crippling superficiality.

In our own historical tradition, dying for a cause is often regarded as a noble sacrifice. What about Gallipoli?

Many IS recruits were born in Britain and Australia—mostly to parents who were integrated and not radicalised.

The emergence of Islamic State and a slew of terrorist attacks was used to legitimise Islamophobia, a deep suspicion or hatred of Muslims, wherever they were and whatever their views—even those (the great majority) who abhorred violence, had sought the protection of an open, stable multicultural society, and feared the radicalisation of their children.

Years of fighting in the Middle East, including in Syria, Afghanistan, and Yemen, contributed to the refugee crisis.

'Government is the enemy': from 'welfare state' to 'market state' (1979–)

The English economist John Maynard Keynes (1883–1946) argued powerfully for government intervention, where appropriate, to deal with crises or market failures, to respond to natural disasters or pandemics, to sequester resources during wartime or for post-war reconstruction, and to provide access to education and health services. The state could intervene to reduce unemployment and provide social security to prevent destitution, using measures such as progressive taxation, adjusting the currency, or going into deficit.

His approach influenced Roosevelt's New Deal in the United States, and was generally adopted in the United States, Britain, Canada, Australia, and New Zealand during World War II and in the decades thereafter, often with bipartisan support.

However, Keynesian economics was sharply attacked by Friedrich von Hayek, Milton Friedman, and the 'Chicago school', who argued that the market—the preference of consumers—was the best allocator of resources, and strongly opposed central planning. They derided the welfare state as 'the nanny state', and opposed any extension of state power which, they argued, would ultimately lead to totalitarianism, as demonstrated by Stalin, Hitler, and Mao.

Margaret Thatcher in Britain, an enthusiast for Hayek, rejected the Keynesian interventionist model, insisting, in 1979, that 'there's no such thing as society'. Ronald Reagan followed in the United States, in 1981, asserting that 'government is not the solution, government is the problem'. The Reagan model was largely adopted by Bill Clinton's Democrats, who deregulated the banks

From the 1980s, Britain and the United States were committed to smaller government, lower taxes, balanced budgets, self-regulation, growth as an end in itself, and an emphasis on individualism, rejecting the concept of 'the public good'. Ethics and fairness were treated as irrelevant. The nation state was replaced by the market state: citizens became 'customers', and health, education, and aged care were defined as 'industries', rather than as essential responsibilities of the state in the human life cycle. Tony Blair retained much of the Thatcherite model in Britain, but in Australia Hawke and Keating skilfully evolved a hybrid, retaining some Keynesian elements such as the 'social wage'.

Progressive income taxes, which had been very high during World War II, and remained high in the post-war decades, were cut dramatically in the US and UK, less so in Australia. This contributed to the widening gap between rich and poor.

The end of communism as an ideology, and the triumph of capitalism and globalisation (1989–)

In 1989, China buried Mao Zedong's ideology (but not his political model) when 'paramount leader' Deng Xiaoping closed the communes, visited the United States, proposed an 'open door' economic policy, and invited multinational corporations to invest. The Tiananmen Square massacre in Beijing demonstrated that the Communist Party retained totalitarian control, and that 'communist' had just become a brand name.

Also in 1989, with the fall of the Berlin Wall, when the communist system in Eastern Europe collapsed and the Soviet Union imploded, only one economic model was left standing in the world: triumphant capitalism, along with the doubtful assumption that it invariably promoted democracy. All values became economic, and all goals material. *Homo sapiens* became *Homo economicus*. 'Left' politics was deemed obsolete, and the lack of any debate on alternative models of society drained politics of idealism, replacing it with a narrow commitment to economic self-advancement.

Francis Fukuyama, who became an inspiration to American neoconservatives with his thesis on 'the end of history', later came to repudiate his admirers. He argued that, through a combination of ignorance and incompetence, they assumed that communism's sudden collapse in 1989 would be a model for an equally sudden collapse of Islamic fundamentalism, after which democracy, American style, introduced by military force, would emerge as a default position in the Middle East.

Globalisation became an achievable goal with the breaking down of tariff barriers and the irresistible expansion of multinational corporations. Local and regional resentment of and suspicion about globalisation was suppressed, only to break out decades later.

'The War on Terror' (2001–)

On 11 September 2001, the twin towers of the World Trade Center in Manhattan were destroyed after two commercial aircraft, hijacked from Boston, crashed into them, leaving 2,996 dead. This attack changed politics in the West beyond recognition. The nineteen suicide hijackers involved were members of al-Qaeda ('the base').

Al-Qaeda's attack was a turning point in modern recent history, marking the international export of jihadist terror and the adoption of elements of a surveillance state in the West. Its leader, Osama bin Laden, coming from a rich and powerful Saudi family, had been inspired by the teachings of Sayyid Qutb.

The attack also shocked and shamed the CIA, FBI, and NSA (National Security Agency), which had shredded its expertise on Islam and the Middle East.

The invasion of Iraq (2003)

The declaration of a 'war on terror' by President George W. Bush, prime ministers Tony Blair and John Howard, and other leaders was an understandable and politically potent reaction to the destruction of the World Trade Center. Ultimately, this led to a military reprisal by the United States, the United Kingdom, Australia, and other members of 'the coalition of the willing' in invading Iraq in 2003. Bush's use of the word 'crusade', was unfortunate.

Military action was directed, bizarrely, as it turned out, at the wrong target. But it was not an accident; it was a deliberate choice.

Saddam Hussein was a brutal and sadistic dictator in Iraq, but his Baathist regime was secular, not jihadist. His elusive 'weapons of mass destruction' (WMD), used as the justification for invasion, were striking examples of Western governments using intelligence material without discrimination or scepticism. However, the US had sold Saddam weapons during his war against Iran (1980–88), so there were some grounds for concern. Oddly, his refusal to provide full disclosure about Iraq's lack of WMD, which proved fatal to him, was that he feared Iran more than the US, and suspected Iran of planning a pre-emptive strike.

Osama bin Laden and al-Qaeda were fundamentalists, deeply religious, mostly Saudis, and antagonists not only of Saddam but of the Saudi monarchy. Saudi Arabia, rich in oil, remained tied to the West, although its religious practices and politics were profoundly antithetical.

The Bali bombings of October 2002, in which 88 Australians were killed as the result of a terrorist attack carried out by members of Jemaah Islamiyah, an Islamist group aligned with al-Qaeda, had a powerful domestic impact.

Reaction to Islamic terrorism led to attacks of senseless violence against Muslims by armed white racial supremacists in the United State, France, New Zealand, and many other countries. Jews were also attacked by racists and some uncomprehending gangs that invoked the name of Hitler and used Nazi emblems.

The reaction to extreme violence undercut the moral justification for multiculturalism, a subject on which contemporary leaders have fallen silent. It also gave the green light to many aspects of the surveillance state, using the newest technological tools, suppression of human rights, including imprisonment without trial, and imposing censorship and secrecy.

The sounds of the dog whistle hang in the air.

After 2001, fears of terrorism and an emphasis on national security and patriotism dethroned reason, elevated irrationality, and debased democratic practice, perhaps permanently. The rule of law was now disposable; truth, evidence, and analysis became marginal, or irrelevant; and the concept of 'the new normal' in the United States and in Australia led to fears that arguing an alternative point of view would be political suicide.

Equal value for votes: equal value for opinions?

We accept the proposition that all votes in an election should be of equal value, irrespective of the extent of experience (or non-experience) of individual voters. But—and here's the problem—does this mean that all opinions are of equal value and should be treated with equal respect, so that vaccinators and anti-vaccinators, for example, should have 50/50 access to the media?

In 1838, in an essay on Jeremy Bentham, John Stuart Mill

coined the term 'fractional truths'. As a practice, it has caught on alarmingly. Something that was true yesterday, and might well be true tomorrow, is not necessarily true today.

In *The Origins of Totalitarianism* (1951), Hannah Arendt wrote:

> In an ever-changing, incomprehensible world the masses had reached the point where they would, at the same time, believe everything and nothing, think that everything was possible and that nothing was true ... Mass propaganda discovered that its audience was ready at all times to believe the worst, no matter how absurd, and did not particularly object to being deceived because it held every statement to be a lie anyhow. The totalitarian mass leaders based their propaganda on the correct psychological assumption that, under such conditions, one could make people believe the most fantastic statements one day, and trust that if the next day they were given irrefutable proof of their falsehood, they would take refuge in cynicism; instead of deserting the leaders who had lied to them, they would protest that they had known all along that the statement was a lie and would admire the leaders for their superior tactical cleverness.

The ideal subject of totalitarian rule is not the convinced Nazi or the convinced Communist, but people for whom the distinction between fact and fiction (i.e., the reality of experience) and the distinction between true and false (i.e., the standards of thought) no longer exist. 'Opinion' is now preferenced over 'evidence', 'feeling' over 'rationality', and science and free enquiry are rejected or discounted. Debate has been not merely infantilised but becomes irrelevant, and language is often debased.

Many prefer to communicate through 'memes': non-verbal, even pre-literate, transmission by gesture, clothing, symbols, and branding.

News as entertainment

Contrary to what might have been expected, with Western communities having unprecedented access to higher education and to information on a massive scale, the public discussion of complex and controversial issues has fallen into an abyss. Facing fierce competition for audience ratings and advertising revenue, the mainstream media—newspapers (with sharply declining readership), radio, and television (facing competition from Netflix, YouTube, and Facebook) have, with some notable exceptions, opted for 'infotainment'. The cult of personality is dominant, based on trivialisation, the infantilisation of debate, and telling audiences exactly what they think they want to hear and see. Investigative reporting, except with crime stories, is out, and as Neil Postman's book title of 1985 suggests, we run the risk of *Amusing Ourselves to Death*. The top-rating programs are gardening, cooking, sport, and 'reality television'.

Public broadcasters such as the ABC and SBS have a chartered responsibility to be fair and balanced, and to provide educational programs, but this does not apply, in practice, to commercial outlets.

Advertising and marketing

Contemporary life is marked by a constant bombardment of advertising messages—to consume, as part of the national mission to keep the GDP rising. So, through social media or television, radio, and magazines, there are appeals, almost demands, to gamble, to consume more food and drink, to buy more hardware, and to create more waste.

Advertisers and promoters are not on oath to tell the truth: we expect to be lied to, and when the lies are exposed, we shrug it off. It's part of the game. Lying becomes normative.

Sport, a central element in Australian life, has always relied on passionate identification, hyperbole, and exaggerated expectations.

Revelations about cruelty in horse racing and dog racing, drugs in football and athletics, corruption in FIFA and the Olympics, and cheating in cricket receive mixed reactions—from denial to disgust—but there is always a PR consultant on hand to massage the story (for a fee). Anybody who called for a strict code of truthfulness would be dismissed, literally, as a spoilsport.

We used to trust churches, banks, insurance companies, aged-care providers, and the police. We know better now. Political parties were never regarded as paragons—but, with compulsory registration and voting, citizens turn up, do their duty, then turn their attention elsewhere.

We don't expect real estate developers or tourism marketers to be models of truth—but we respond to voices that appeal to our inner selves. Journalists, too, can be great spinners.

It is no coincidence that Donald Trump was a real estate developer and media performer, with a chequered career; Boris Johnson, a journalist who habitually made things up as he went along; and Scott Morrison, a marketing executive whose reputation was made with Tourism Australia's campaign slogan, 'Where the bloody hell are you?' This slogan came to haunt Morrison when he departed for a Hawaiian holiday at the height of the bushfire emergency in December 2019.

So in the maelstrom of marketing ('Have we got a deal for you?'), truth, honesty, and evidence become disposable.

Hate media

Talkback radio was introduced in Australia in April 1967. I should declare an interest: I hosted the first talkback radio program in Melbourne. However, this innovation soon became unattractive and dangerous. In my program, I sought to educate, and—long before the Internet—had my own database. When generating conversation about subjects such as Vietnam, or drug abuse or suicide, I could almost always justify any material presented because I had the latest data at hand. My guests, ranging from Sir Robert Menzies to Bob Santamaria to Dr Jim Cairns, reflected a variety of viewpoints, and there were no shouting matches. Perhaps surprisingly, my program led its competitors in the ratings.

Talkback radio began even earlier in the United States, but after 1988 it was weaponised by Rush Limbaugh and his many imitators, who captured their audience by providing a series of shocks, with highly selective material laced with fear, hate, and nostalgia.

The American 'shock jocks' played an important role in promoting the Tea Party movement and pushing the Republican Party towards embracing populist nativism.

Later Australian talkback hosts—notably John Laws, Alan Jones, and Ray Hadley—were shock jocks, following the Limbaugh model. Their goal was to provide a mixture of entertainment, outrage, shock, and awe, and to provide an opportunity for favourites to talk at length, without any sharp questioning. The truth of any proposition was inconsequential. Talkback radio became John Howard's outlet of choice, and he used it skilfully. So did Tony Abbott, Peter Dutton, and Scott Morrison. Malcolm Turnbull disliked talkback radio, and its shock jocks laboured mightily to bring him down.

The shock jocks will say anything, so long as it rates. Heavy damages ($3.4 million) awarded in 2018 against Alan Jones for libelling the Wagner family in Queensland had no effect on the loyalty of his (mostly ageing) followers. But the Jones devotees, although a minority, had a powerful influence when their votes were needed in a close election. So politicians tended to be very respectful when Mr Jones called, and they rarely challenged him.

Jones (until his retirement) and Hadley have had a large audience in New South Wales and Queensland, but have not been broadcast in Victoria.

Shock jocks claim to be omniscient on every subject they tackle, and their every pronouncement is categorical—there are no nuances or attempts to grasp an alternative point of view, or even to discuss evidence. Two styles are used: obsequious and bullying.

They are essentially fundamentalists. The need for a target is central.

Shock jocks are particularly shrill on climate change (denialist), asylum seekers ('We're not tough enough!'), feminism, Indigenous causes, especially native title, 'the history wars', multiculturalism, 'political correctness' (a useful term, never clearly defined), crimes attributed to migrants, the Greens (even more than the ALP), and Islam. Misogyny is a common theme, and attacks on Julia Gillard and Jacinda Ardern (although not Margaret Thatcher, as I recall) were shrill, even unhinged.

In the United States, supporters of Donald Trump were likely to watch Fox News exclusively because every report reinforced their preconceptions, leaving them unaware of alternative information or opinions.

Opponents of Trump preferred the television news coverage on CNN, MSNBC, and, to a lesser extent, CBS.

'The Medium is the Massage'

The Canadian philosopher and media analyst Marshall McLuhan argued, prophetically, in *The Medium is the Massage: an inventory of effects* (1967) that it is the medium itself—rather than information, data, evidence, analysis, and reason—which 'massages' audiences and conditions the way people understand the world and they react to it:

> All media work us over completely. They are so pervasive in their personal, political, economic, aesthetic, psychological, moral, ethical and social consequences, they leave no part of us untouched, unaffected, unaltered.

The media are extensions of our human senses—and we look to them not to be challenged, but to reinforce our view of ourselves in an Age of Anxiety, marked by an unprecedented rate of change.

McLuhan died in 1980. I doubt if Donald Trump ever read McLuhan, but his media career as host and producer of the reality show *The Apprentice*, from 2003 to 2015, demonstrates how a powerful media presence can capture a significant audience share.

After Trump was inaugurated as president of the United States in 2017, having never had any experience in government, he operated, almost entirely, in and through the media. Every morning, he picked up his agenda for the day from Fox News, especially the program *Fox & Friends*. However, when Fox News ventured mild criticism about his recommendation to use hydroxychloroquine for COVID-19, Trump began endorsing One America News Network (OAN), which was totally committed to his causes. He notoriously discounted official briefings, relying on his own 'intuition'.

He could not have been accused of being a workaholic. His activity was constructed around media events—press conferences, campaign rallies, tweeting, stirring up controversies, pursuing grievances. He was enraged by the mildest criticism, and routinely denounced investigative journalists as 'enemies of the people', a French Revolutionary touch!

The triumph of populism (2016)

The year 2016 was marked by three troubling international events: in May, the election of Rodrigo Duterte, a specialist in extrajudicial killings, as president of the Philippines; in June, the referendum vote in the United Kingdom to leave the European Union ('Brexit'); and in November, the defeat of Hillary Clinton by Donald Trump, each following campaigns that made a strong appeal to populism, fear, and prejudice.

Reliance on 'spin'

Managing the media—and especially massaging perceptions—is a central preoccupation of modern governments, institutions, businesses, and celebrities, including media stars and members of royal families.

The public relations industry has become an integral part of modern government and business, massaging information and providing appropriate mantras that can be repeated, endlessly, as part of a 24-hour news cycle.

Governments rely on spin, employing spin doctors to craft explanations, plausible or otherwise, for every failed policy. The hollowing out of expertise in government departments, and the ever-greater reliance on ministerial staff to develop politically calibrated 'policy', has weakened political accountability.

Both government and media are inextricably linked to lobbyists or pressure groups, who are happy to suggest what should be said, or published, and when.

In the United States, the National Rifle Association (NRA) has been exceptionally powerful in blocking even minor reforms to gun ownership, despite polling indicating very high levels of public support for restricting access. For every 100 people in the US, there are 120 guns, compared to 35 in Canada, fourteen in Australia, three in the United Kingdom, and less than one in Japan. In 2017, gun-related homicides in the US totalled more than 14,500; Japan had just three.

After each mass shooting in the United States, the spin formula was faithfully repeated by President Trump and others in thrall to the NRA: 'This is not the time to debate gun ownership. This is a time for grieving, and our thoughts and prayers go out to the victims and their families. This killing was the action of a disturbed individual and if only teachers/doctors/sports fans had been carrying guns they could have protected themselves and saved many lives.'

This formula was faithfully adapted by Prime Minister Scott Morrison in response to the unprecedented bushfires in November–December 2019: 'This is not the time for a political debate about global warming. Our priority is with the victims, and our thoughts and prayers go out to them and their families.'

It has to be said in John Howard's favour that after the appalling massacre at Port Arthur on 28–29 April 1996, leaving 35 dead and 28 wounded, he did *not* say: 'This is not the time to talk about gun control. Our thoughts and prayers are with the victims and their families, and that's what we should be concentrating on now.' He took strong and effective (and in some areas unpopular) action, admittedly with opposition support, to legislate for tough gun-control measures. He was courageously supported by Tim Fischer, the leader of the National Party, whose rural base used large numbers of military-style weapons.

It turns out that it is never the right time to debate either gun control in the US or global warming in Australia.

Scott Morrison and Mathias Cormann make a point of never answering direct questions, using such formulas as, 'That's only of interest inside the Canberra bubble', and (when scandals are exposed), 'I don't answer questions based on gossip', or 'I'm not going to comment on an issue that the Labor Party has raised.' Morrison has an unblemished record of never having been around when any policy debacle was cooked up. He invariably fails to take responsibility for errors, failures, misstatements, or exaggerations. Nobody told him, he was never shown the file, or he was on leave. His hands are spotless.

If a gold medal was awarded for avoiding direct answers to questions, Morrison would be an obvious winner. In Australia and elsewhere, ministers employ minders in their offices as a protective firewall to enable them, when caught out, to plead plausible deniability. Both major parties are adamant that minders must be protected by 'executive privilege', so that they cannot be called to give evidence, let alone be subject to cross-examination, by parliamentary committees or integrity commissions.

Cabinet ministers are not necessarily lying when they claim not to understand why a policy failed: they may simply not be aware. They can say, 'I never thought to ask' and that their minders had not wanted to tell them.

Redacting documents, especially on a commercial-in-confidence basis, occurs on an almost industrial scale. There are 20–30 classifications that can be applied to documents, even innocuous ones, in order to defeat Freedom of Information (FOI) requests. Even some classifications are themselves regarded as secret.

Australia's public service had a long and proud tradition of giving frank and fearless advice, based on an examination of evidence, often telling ministers what they do not want to hear, but helping them to avoid catastrophe.

Now the Australian public service is told that its future role is simply to carry out instructions by a minister: no independent, objective, expert advice will be sought.

As Adrian Wooldridge wrote in *The Economist* (7 December 2019), Boris Johnson 'is in many ways the ideal politician for a post-truth age, because nobody expects him to keep his word. He exists in a world of us-versus-them and of emotion rather than reason, a world in which cheering people up is more important than depressing them with facts.'

The same description could be applied to the Australian prime minister, Scott Morrison. It was no accident that Johnson's spin doctors included the Australians Sir Lynton Crosby and Isaac Levido.

The US model of campaigning through the media proved very attractive to Coalition politicians in Australia, and Donald Trump, in particular, attracted the sincerest form of flattery.

Mussolini used to rail about 'draining the swamp'. Trump adopted the term in 2016 without necessarily knowing who Mussolini was, and now Morrison uses the metaphor to describe Canberra.

Morrison told his party room in December 2019: 'We're winning so much you'll be sick of it by next year', which was an obvious riff on Trump's notorious campaign boast in 2016: 'We gonna win so much you may even get tired of winning and you'll say please, please, Mr President, it's too much winning.'

As with Trump, Morrison's campaigns were targeted to electorates with low incomes, low numbers of graduates, and who were predominantly Christian.

Nativism, immigration, and punishing refugees

'Nativism' has been defined as an ideological and/or political movement in which an existing settler society, such as the United States or Australia, reacts against migrants from different races, languages, religious, and cultures. In the United States, the Native American Party—generally called the 'Know Nothings'—founded in the 1840s, was anti-Catholic, and hostile to large numbers of Irish emigrants. Former president Millard Fillmore ran as its candidate in the 1856 election, polling 21.5 per cent of the vote.

The creation of the Ku Klux Klan after the US Civil War, and its revivals in 1915 and 1946, became a prototype for European fascism, and its second iteration was studied carefully by the Nazis, with its emphasis on race and the use of violence, uniforms, weaponry, and patriotic appeals.

There are significant nativist elements in contemporary politics. Refugees have been criminalised, with 'asylum seekers' being redefined as 'illegals' and attacked as endangering national security. Before the coronavirus hit, immigration continued at high levels because workers were needed for low-paid jobs that citizens were reluctant to take on.

Immigration, characterised as a threat to job security, became a hot-button issue linking extremists and moderates, the hard right and the old left. Blue-collar workers in traditional industries in regions with declining prospects in a global economy saw migrants in general, and refugees in particular, as a threat.

Hostility to immigration was the common element in the United Kingdom's 2016 referendum on membership of the European Union, where low-income areas in the Midlands and north of England voted overwhelmingly for 'Brexit'; in Donald Trump's victory in the 2016 presidential election, using slogans such as 'Build the Wall' (a physical barrier against Mexican and other Hispanic immigrants) and 'Make America Great Again', with 'again' harking back to an illusory 'golden age' of white hegemony; and in France, where the 'gilets jaunes' ('yellow vests') movement had two wings— left (economic) and right (racist).

The immigration/refugee issue changed the politics of Hungary, Poland, the Czech Republic, Italy, Belgium, and Australia, with the rise of Pauline Hanson's One Nation Party, Clive Palmer's United Australia Party, and some hard-right fringe groups.

Lying becomes normative

When politicians are no longer constrained by a moral obligation to tell the truth, lying becomes standard practice.

In the infamous 'children overboard' incident of October 2001, Liberal defence minister Peter Reith alleged that asylum seekers had thrown their own children into the sea as a strategy, so that the navy would have to rescue them and that this might lead to parents being given passage to Australia. It proved helpful in the election, although Air Marshal Angus Houston, then acting chief of the defence force, made it clear that the allegations were not true.

When concerns were raised about the killing by a policeman of a refugee detained on Manus Island in April 2017, Peter Dutton, as immigration minister, insisted that the incident had not occurred, and kept insisting until the police chief of Papua-New Guinea confirmed the killing and the policeman was charged with manslaughter.

Prime Minister Scott Morrison took a gaggle of journalists to a re-opened detention centre on Christmas Island in April 2019, asserting that allowing refugees access to medical treatment in Australia (known as 'medevac') would lead to an upsurge in the number of refugee boat arrivals. When this did not happen, he made no apology and, within weeks, the 2019–20 budget provided for the facility to be closed.

In 2017, Malcolm Turnbull, as prime minister, and Scott Morrison, as treasurer, repeatedly refused to set up a royal commission into banks and financial services, insisting that the existing system of self-regulation and regulatory agencies was enough to guarantee the integrity of the system. They voted against a royal commission 26 times, and spoke against it even more often. Then they were forced to act, when even the banks agreed that a royal commission had become necessary, and appointed Kenneth Hayne, a former High Court justice, as commissioner. Evidence disclosed widespread corruption and illegality in the banks and financial-services industry, and Hayne's report was scathing. (Westpac, for instance, admitted having committed 23 million breaches of the money-laundering laws.) Turnbull and Morrison then claimed credit for having set up the royal commission, and held themselves out as crusaders for reform.

The ALP's reputation has been damaged by habitual lying or evasion about the sources of their campaign funding, many associated with Chinese influence, the gaming industry, and property developers. It had been shifty, to put it mildly, in denying plans to sell off the Commonwealth Bank and Qantas in 1993, but could always claim the excuse of 'commercial in confidence' to keep the selling price up for both.

The ALP's hyperbole in its 'Mediscare' campaigning in 2016 was taken as justifying the Coalition's right to lie *ad lib* in 2019.

All major parties have murky and secretive internal processes, involving recruitment, funding, and candidate selection (or removal).

Social change: work, language, and culture

In Britain and Australia, Labour/Labor parties were weakened by a contracting base. The labour force was changing dramatically, with more self-employed contractors, fewer workers in trade unions, increasing employment in services, especially nursing and teaching, and some technological displacement. Immigration *per se* was not a major factor in job competition, but it was easy to generate a perception to the contrary.

New issues on the political and social agenda were handled awkwardly by both parties.

Language and culture were sensitive areas that could be exploited by lying on a heroic scale. Australia, like the United States and Great Britain, was both multicultural and monolingual, so there were acute communication problems relating to education, employment, health, and alienation.

It was a classic example of looking for a simple explanation for a complex problem. As the acerbic American journalist and critic H.L. Mencken wrote: 'There is always a well-known solution to every human problem—neat, plausible, and wrong.'

The revival of religion as a political force

While the mainstream religions are in decline and have come under repeated attacks, there has been a significant, and generally unexpected, growth in fundamentalist or charismatic churches, whose numbers worldwide have been estimated at 500 million.

Pentecostals have had a dramatic increase in their numbers in the US and Australia.

The teaching of evolution in schools, and the legalising of same-sex marriage, access to abortion, no-fault divorce, and, in a few areas, 'dying with dignity' is generally accepted by the broad community, but an affronted minority of dissenters has become politically active and determined to either change the laws or prevent similar changes being adopted. Because their beliefs are founded in faith, including the inerrancy of the Bible, miracles, and the rejection of much science, fundamentalists are not shaken by, or even interested in, evidence. This makes it difficult to engage in reasoned discussion with them. However, adherents are eager to exert political influences over causes they hold dear.

Among mainstream religions, there has been a significant shift in the way Catholics vote. Until 1964, in the United States, Catholics voted overwhelmingly for Democratic candidates for president, but their votes are now more evenly divided. Exit polls indicate that in 2016 Trump won the Catholic vote by 52 per cent to 45 per cent. The Democrats cannot win without the Catholic vote, and hence must appeal to them. This raises serious problems with its secular supporters, who resent making concessions when the Church is under attack and is in retreat globally. It is no accident that Joe Biden is a Catholic.

The impact of religion in contemporary politics is discussed in Chapter 8.

The collapse of trust

The decline in trust in institutions has become almost universal in Western democracies in recent decades. In the case of the US, there is a disturbing factor—the high level of trust in organisations that operate under a disciplined chain of command, are heavily armed, and are operationally secretive.

Annual polling by the Gallup organisation from 1973, and more recently by the respected Pew Research Center and Forbes-Statista, indicates a high degree of convergence about what institutions Americans have 'a great deal/quite a lot' of trust in:

- the military: 74 per cent;
- small business: 67 per cent;
- the police: 54 per cent;
- the Church/organised religion: 38 per cent;
- the presidency: 37 per cent;
- the Supreme Court: 37 per cent;
- the medical system: 36 per cent;
- public schools: 29 per cent;
- newspapers: 23 per cent; and
- Congress: 11 per cent.

Trust in universities and scientific institutions was not polled.

These rankings help to explain why the reaction of many Americans to the coronavirus threat was to buy another gun and refuse to wear a protective mask.

In Australia there has been a massive collapse of trust, for a variety of reasons, in the country's major institutions, including parliaments, political parties, churches, schools, banks, insurance, aged-care institutions, prisons, retailers, the media, some elements of the armed forces and police, some elements of sport—the national religion—electricity retailers, and the gambling industry.

The Australian National University has conducted surveys of voters for each election since 1987, first under Professor Don Aitkin, and then under the direction of Professor Ian McAllister. While such polls are not infallible, they are a valuable indication of trends in Australian political opinion.

The 2019 survey, of 2,179 voters, found that 41 per cent were not satisfied with democracy in Australia; only 31 per cent 'took a great deal of interest' in that year's election campaign; only 25 per cent thought that people in government could be trusted; only 29 per cent followed party 'How to Vote' cards; and only 17 per cent of Coalition voters and 14 per cent of ALP voters regarded themselves as 'lifelong party supporters'. Drawing on earlier surveys, McAllister found that in 1967, 72 per cent had indicated that they always voted for the same party, but this had fallen to 39 per cent in 2019. The survey found that 30 per cent of voters supported different parties for the Senate and the House of Representatives, opting, in effect, for 'Stop' and 'Go'.

The 'Post-truth era'

Xenophobia, racism, hostility to migrants/refugees, and misogyny have become dominant factors for an alienated minority. However, minorities can exert great power in a close election. While it is rare for them to win seats in the House of Representatives, Australia's system of preferential voting means that their votes determine who wins the election.

We can call this 'the Post-truth era'. In this era, despite Australians, Americans, and Britons having incomparably better access to higher education than previous generations, and with the world's intellectual capital instantly available in a handheld device, the quality of public discourse on major issues has plummeted. The truth of a proposition is no longer central, or even relevant. In public affairs, in the age of spin, with the enormous power generated by modern communication techniques and through social media, every proposition can be targeted to millions of people, taking account of their fears, prejudices, desires, ambitions, and dislikes—even hatreds.

And consumers are themselves consumed—and their personal histories mined for exploitation.

It may surprise Australians aged under 50 to learn that there was a time, only decades ago, when social, political, and economic changes were shaped by reasoned debate, largely dependent on the examination of evidence; ideas were fiercely contested; data was challenged, checked, and recalculated; and the penalties for outright lying could be severe.

More than 250 years ago, Europe and North America were transformed by a dramatic new approach to collecting information, analysing it, and following the consequences. We call it the Enlightenment, and Australia's European settlement was a direct consequence of it.

CHAPTER THREE

Overturning the Enlightenment

The eighteenth-century intellectual movement known as 'the Enlightenment' shaped Europe and North America—and Australia, too—for 200 years. Of growing significance from the 1730s onwards, the Enlightenment was essentially an extension and expansion of the scientific revolution of the seventeenth century (headed by Kepler, Galileo, Bacon, Descartes, Newton, Leibniz, and Spinoza), often described as the 'Age of Reason'.

Its starting point was the observation of phenomena, and their explanation, followed by dissemination and the promotion of practical applications.

Enlightenment values (sometimes called 'the Enlightenment project') are currently under attack. Understanding what that proposition means requires a definition of 'enlightenment'.

The German philosopher Immanuel Kant, in a famous essay 'What is Enlightenment?' (1784), defined it as 'man's emergence from his self-imposed immaturity. Immaturity is the inability to use one's understanding without guidance from another.' The Enlightenment could be summarised as '*Sapere aude*!' ('Dare to know!')

The term 'Enlightenment' in English, '*Siecles des lumières*' in French, '*Aufklärung*' in German, '*Illuminismo*' in Italian, and '*Ilustración*' in Spanish was used retrospectively as a convenient description of what had happened. This was also the case with the Agricultural and Industrial Revolutions.

Unfortunately, from the 1970s, the word 'Enlightenment' lost its historic meaning and was misapplied in the context of 'the Age of Aquarius' or 'the New Age', self-awareness cults, feng shui, the transforming power of crystals, mysticism, karma, and holistic health.

The exploration of Australia and its settlement/occupation/invasion from 1788 was inextricably linked with the Enlightenment and the Agricultural and Industrial Revolutions of the 1780s that followed from it.

James Cook's exploratory expedition to the Pacific (1768–71) to observe the transit of Venus (1769), and then, after circumnavigating New Zealand, to chart the east coast of *Terra Australis Incognita*, the barely understood Great Southern Land (1770), was a characteristic Enlightenment venture.

Cook's retinue included Joseph Banks, an amateur scientist and patron of learning, who partly funded the expedition; the Swedish botanist Daniel Solander, a disciple of Linnaeus, the preeminent taxonomist; a Finnish botanist, Herman Spöring; and the extraordinary botanical artist Sydney Parkinson. Banks brought back 30,000 botanical specimens.

In a second voyage (1772–75), Cook explored the southern oceans, returned to New Zealand (but not Australia), crossed the Antarctic Circle (1773), and reached 71°10' S.

In the United Kingdom, the Agricultural Revolution, including the enclosure of common lands, led to dramatic increases in production, with many families of tenant farmers forced off the land. The significant movement of population to cities, looking for work, resulted in squalid living conditions, an upsurge in crime, and the need to find some faraway place to transport convicts instead of hanging them or chaining them in prison hulks. And since 1776 the American colonies had no longer been available for transportees.

Banks proposed the settlement of what was originally known as 'New Holland', later renamed New South Wales.

From January 1788, when the eleven ships in the First Fleet arrived with Arthur Phillip as governor, the building blocks of the British annexation and occupation of the east coast of the Australian continent were convictism and Aboriginal dispossession. Some aspects of 'convictism' could be described as enlightened. At Port Arthur, regarded as a very harsh prison, nutrition, health, and the average age at death was superior to the rest of Van Diemen's Land. There was a well-used library, a cricket pitch, and a garden. An iron foundry and ship-building both required skilled labour.

Aboriginal dispossession was the dark side of European settlement. White invaders (or 'settlers', depending on one's position in the 'history wars'), in addition to dispossessing Aborigines and Torres Strait Islanders from their land, murdered many thousands, exposed them to new diseases, which killed far more, and followed this with the destruction of Aboriginal cultures and languages, condescension, the breaking-up of families, marginalisation, and neglect.

Above it all was what the anthropologist W.E.H. Stanner called 'the great Australian silence', a form of cultural amnesia, buttressed by ignorance, an *apartheid* more psychological than physical or political. It lasted for almost 200 years.

But early British and French explorers, and administrators, were fascinated by Australia's distinctive flora and fauna, which they collected assiduously, sending many thousands of specimens back to collections in Europe. (The most enthusiastic collector of Australian fauna and flora was the Empress Josephine, the wife of Napoléon Bonaparte, who, at her Chateau de Malmaison outside Paris, kept kangaroos and emus, and bred Australian black swans.)

The relatively new science of anthropology meant that 'Indigenes' attracted intense interest—at first involving social interaction, but soon relegated to the status of 'specimens', dead or alive. Within a generation, Aboriginal people were pushed to the margins, their populations slashed by the introduction of diseases to which settlers had become immune. Then followed a long series of frontier wars: an unequal, brutal, and unedifying struggle.

Early governors in New South Wales were seriously committed to the Enlightenment, and especially interested in science. They experimented with new approaches to organising the settlement on rational lines, importing innovations in farming and manufacturing.

The central element in the Enlightenment was scepticism about an anthropomorphic God who intervened in the daily lives of humans.

This meant that the Bible was not the only source of wisdom and knowledge; the concept of miracles was rejected; faith was challenged by reason; 'soul' was reconceived as 'mind'; church and state were separated; and the divine right of kings and queens was rejected. It followed that every person should have the right to pursue truth in his/her own way; blasphemy/heresy would no longer be prosecuted; education should be a state responsibility and not a church monopoly; and Jews should have full rights as citizens, and all religions should be tolerated. 'Natural law', derived from observation, could be applied to human affairs.

The great writer Voltaire, the pen name adopted by François-Marie Arouet (1694–1778), was the most passionate campaigner for religious toleration.

The contemporary Israeli historian and public intellectual Yuval Noah Harari argues that, 'Liberalism took a radical step in denying all cosmic dramas, but then recreated the drama within the human being—the universe has no plot, so it is up to us humans to create a plot, and this is our vocation and the meaning of our life.'[1]

Science provided explanations of phenomena—eclipses, meteorites, lightning, thunder, earthquakes, volcanoes, flood, climate, disease, famine—that had once been regarded as a demonstration of God's anger (like Noah's flood).

A remarkable group of scientists, including Benjamin Franklin in America, Mikhail Lomonosov in Russia, Henry Cavendish and Joseph Priestley in England, and Antoine Lavoisier in France, promoted the application of scientific method and, with Carolus Linnaeus in Sweden and Joseph Banks in London, the classification and publication of data. This included the isolation and naming of elements and of species of flora and fauna. Lavoisier also promoted the decimal system and uniform measurements. Franklin, Alessandro Volta, and Luigi Galvani experimented with electricity, and Horace de Saussure captured solar energy in 'hot boxes'.

There was intense interest in astronomy, chemistry, physics, geology, palaeontology—and in determining the age and mass of the Earth, and the size of the universe. William Herschel discovered the planet Uranus (1781).

A zeal for exploration was intense: James Cook, Louis de Bougainville, Jean-François La Pérouse, and Nicolas Baudin explored the Pacific, and Alexander von Humboldt spent years in South America. Curiosity about Antarctica was promoted by Lomonosov, and Cook ventured far South. Australia's unique biota stimulated consuming interest.

There were significant medical advances, including vaccination against smallpox (Edward Jenner), and recognition of the need to eat citrus fruit, and fresh vegetables to prevent scurvy (James Lind and Cook). Rational treatment of the insane was pioneered by Philippe Pinel.

The social sciences, education, and philosophy were also fundamental to the Enlightenment.

The concept of 'checks and balances'/'division of powers' in government, essentially a limitation on the executive, which was pioneered by John Locke and Charles Louis de Montesquieu, was followed up by Thomas Paine, Thomas Jefferson, James Madison in America, and, to a degree, Edmund Burke in England.

In their Declaration of Independence (1776), the thirteen American colonies proclaimed the goals of 'life, liberty and the pursuit of happiness' as 'an inalienable right', expressly for white, male, English-speaking settlers, but the document was silent about women and slaves. Quasi-democratic institutions were established in the United States' Constitution (1789), influenced by Locke, Montesquieu, and Rousseau, and the Bill of Rights (1791) contained some innovations, especially on freedom of expression (although it was bad on guns, even if the intent was to legitimise the arming of militias rather than individuals.)

Jean-Jacques Rousseau, who died in the same year as Voltaire (1778), was an intensely subjective contrarian who challenged many Enlightenment precepts, and with his cult of emotion (*sensibilité*) became the most important forerunner of Romanticism. His concept of child-centred education, set out in *Émile* (1762), was revolutionary. *The Social Contract* (1762) expanded the ideas of Hugo Grotius and Thomas Hobbes, arguing that political authority depends on the consent of the governed, who surrender some rights in order to secure social stability and mutual protection. He also promoted the concept of 'the general will', which became a central theme of the French Revolution.

The Milanese economist Cesare Beccaria, in his pioneering work *On Crimes and Punishment* (1764), demonstrated that the greatest deterrent to crime was the certainty of apprehension and conviction, not the severity of punishment, and argued that the death penalty 'was neither necessary nor useful'. Beccaria also used the phrase 'the greatest good for the greatest number' (*'la massima felicità divisa nel maggior numero'*), a concept popularised in English by Francis Hutcheson and the utilitarian Jeremy Bentham.

There were also moves to end torture and growing opposition to the slave trade.

The philosopher David Hume promoted scepticism about dogma and promoted openness to new ideas. He recognised that pure reason was insufficient to explain human behaviour, and emphasised the need to understand irrationality in decision-making. (This idea was expanded in *Thinking Fast and Slow* (2011) by the Nobel prize-winning psychologist Daniel Kahneman, developing themes worked out with Amos Tversky.)

An analytical approach to history was promoted by Giambattista Vico, Edward Gibbon, and William Robertson. Adam Smith attempted to explain economics on a rational basis. William Jones proposed the hypothesis of an Indo-European language from which most of Europe's languages were derived. Recognition of tensions between the universal and the local, for example with language and custom, marked the beginning of what we now call sociology.

There was a heavy emphasis on the collection and dissemination of information.

Ephraim Chambers' *Cyclopædia: or, An Universal Dictionary of Arts and Sciences* (1728), published in London, was an ambitious precursor of the Enlightenment, and its model was adopted in France.

The great *Encyclopédie* (1751–72), published in seventeen volumes of text with eleven volumes of illustrations, contained 72,000 entries by 140 authors in 16,288 pages. Edited in Paris by Denis Diderot and Jean-Baptiste d'Alembert, it challenged the authority of scripture and revelation, and played an important role in setting the scene for the French Revolution (1789–99). The *Encyclopædia Britannica,* first published in Edinburgh in three volumes (1768–71), remained in print for 244 years.

With increased literacy came the development of newspapers and journals, the relaxation of censorship, and promotion of the arts and sciences—in theatres, concert halls, and public lectures. Early experiments in democracy were attempted in Switzerland, Sweden, the United States, and France. Statistics and evidence were marshalled to argue for change, and the role of the expert was taken seriously. It was assumed that shared data might lead to shared experience.

There were attempts to reach out to the universal, to explain complexity and to promote the concept of the public sphere.

The central concepts of curiosity, openness, and a willingness to seek expert advice, and to rely on evidence are now, 200–250 years later, hanging by a thread.

England had an ambiguous relationship with the European Enlightenment, both intellectually and in the way its institutions operated. King Charles I had been executed in 1649, and the concept of 'divine right' went with him. Experimentation with republicanism followed in Oliver Cromwell's Commonwealth. Breaking with the universal church in Rome had happened even earlier, in 1534. The magnates who ruled Great Britain in the eighteenth century were pragmatic and secular.

By 1689, after the so-called 'Glorious Revolution' enshrined parliamentary supremacy, Great Britain had a Bill of Rights—something that Australia, after 230 years of European occupation, has failed to achieve and shows little interest in pursuing.

The last veto of an Act of Parliament by a British sovereign was in 1708 (and even then, Queen Anne acted on ministerial advice), and it can be safely assumed that the royal veto is now defunct. However, in the Commonwealth of Australia Constitution (1901), the royal veto is explicitly preserved in sections 58 and 59.

Somehow, we cannot face the idea of leaving home.

Curiously, the three great English (or Anglo-Irish) literary masterpieces of the eighteenth century, Samuel Johnson's *A Dictionary of the English Language* (1755), Laurence Sterne's anarchic comic novel *The Life and Opinions of Tristram Shandy, Gentleman* (1759–67), and Edward Gibbon's *The Decline and Fall of the Roman Empire* (1776–88) were never regarded as being part of the Enlightenment, although all three were influential in Europe.

Scotland, which had more universities (four) in the eighteenth century than England (two), despite its smaller population, was much more deeply involved in the Enlightenment. In Edinburgh and Glasgow there was intense intellectual speculation and the dissemination of knowledge—by Francis Hutcheson, William Robertson, James Lind, David Hume, Adam Smith, James Hutton, Joseph Black, Lord Morton, Lord Monboddo, Lord Kames, James Watt, and the publishers of *Encyclopædia Britannica*.

It is difficult to settle on a closing date for the Enlightenment. The outbreak of the French Revolution in 1789 is often regarded as the beginning of Romanticism, which was, at least in part, a convulsive reaction against the Enlightenment.

However, Enlightenment values survived and enjoyed a resurgence after the end of World War II, and even more after the Cold War ended in 1989.

'The Enlightened Despots'

In Europe, the principles of the Enlightenment were promoted and applied by a remarkable group of rulers—exact contemporaries— attracted by the idea of reforms that consolidated their own power and weakened the influence of religion. They were very widely read.

These included Frederick the Great of Prussia; Maria Theresa of the Holy Roman Empire, and her sons, the Emperor Joseph

II and Grand Duke Leopold of Tuscany, and daughter, Maria Carolina of Naples and Sicily (through her husband); Catherine the Great of Russia; Carlos III of Spain; José I of Portugal (through his minister, Pombal); Gustav III of Sweden; and Christian VIII of Denmark (through his minister, Struensee). Frederick, an able military strategist, was a flautist and composer who invited J.S. Bach to Potsdam.

Louis XV and XVI of France had enlightened people around them, but were resistant to major changes. George III of Great Britain, passionately opposed to Catholic emancipation, was a great reader and book collector (with 65,000 volumes), and played the harpsichord.

Emperor Joseph II was an anomaly—an autodidact who lacked sympathy with the French *philosophes*, essentially a social engineer, strongly secular, widely travelled, but unable to generate sympathy for his ambitious plans for change. His proposal to impose German as the uniform language of his polyglot empire was rational from a utilitarian perspective, but certain to inflame Slavs, Hungarians, Italians, and Flemings, who comprised the majority of its population. He promoted state education and infuriated the church by stripping monasteries of property and by making a utilitarian attempt to abolish coffin burials as a waste of resources.

Joseph II, his brother, Leopold, and Catherine the Great all abolished capital punishment and torture for civil crimes in their domains.

Many of the *philosophes* had international reputations, and were taken up by the enlightened despots. Voltaire went to live (not very happily) at Frederick's court, and Diderot was Catherine the Great's (mostly absentee) librarian.

After Edward Jenner proposed vaccination as a prophylactic for smallpox, Catherine was vaccinated herself in 1768 to encourage her subjects to accept the novel treatment.

Critics of the Enlightenment

Leftist critics of the Enlightenment argued that the replacement of church authority by the secular state contributed to authoritarian rule, that quasi-scientific principles were used to justify the racial supremacy of Europeans and the expansion of imperialism, and that its mechanistic worldview was the intellectual basis for capitalism and the exploitation of human and physical resources. As A.C. Grayling points out (*The History of Philosophy*, p. 273), antagonists blamed the Enlightenment for 'its supposed superficial rationalism, foolish optimism, and irresponsible Utopianism', citing the Terror in the French Revolution, the Napoleonic wars, and both Stalinism and fascism. Contemporary critics, especially in the United States, and in Australia, see the Enlightenment as an attack on 'family values'.

But the horrors of the twentieth century were shaped far more by the very factors that the Enlightenment argued *against*: nationalism, appeals to force and emotion, and politics as theatre. ('A preference for rallies over debates.')

'Romanticism' was partly a reaction against the restrained neo-classicism of the Enlightenment.

The methodology of the Enlightenment was inextricably linked with the eighteenth-century's Agricultural Revolution and the Industrial Revolution. Collectively, they produced models of evidence-based problem-solving that transformed the world, well until the twenty-first century.

The Agricultural Revolution

The eighteenth-century Agricultural Revolution originated in England, Flanders (modern Belgium), and the Netherlands, but was gradually adopted throughout Europe, North America, and—before long—in Australia.

Essentially, small-scale subsistence farming, based on the collective use of common land, which had evolved from feudalism and sustained village life, was replaced by large-scale agriculture as an industry, shaped by capitalism. From the sixteenth century, common lands had been gradually enclosed and fenced off, and displaced farmers or labourers had to find other work.

With the enclosure movement, crops were grown, or animals grazed, in larger units of land than had been traditional practice. Seed was sown in rows, instead of being broadcast. Three-year, and later four-year crop rotation (wheat, barley, turnips, clover) regenerated the soil. Potatoes and maize were introduced as staples. Experimentation with grains and vegetables led to far greater output. The selective breeding of cattle, horses, and sheep, and the availability of more fodder led to dramatic increases in the size and weight of farmed animals. (The famous 'Durham ox' probably weighed 1,300 kilograms.) New farming equipment, including the seed drill and iron ploughs, drawn by animals, led to far greater production with fewer workers. There were changes in watering, drainage, and aerating the soil. All these changes were based on close observation, trial and error, countless experimentation, careful recording—and the dissemination of knowledge.

It was also recognised that free labourers were likely to be more productive than serfs.

Outstanding figures in the Agricultural Revolution included Jethro Tull (not the musician), Thomas Coke, Robert Bakewell, Charles Townshend, and Arthur Young.

Robert Walpole, the longest-serving British prime minister of the eighteenth century, was a great 'improving' landowner. King George III ('Farmer George') was a proponent of agricultural reform, and (as 'Ralph Robinson') wrote knowledgeably about it.

The Agricultural Revolution caused social disruption, ending access to common land, driving most subsistence farmers away from holdings that their families had tilled for generations. This led to a massive population influx to cities and towns.

For the first time in human history, agriculture ceased to be the dominant employer. Agricultural employment has been estimated to have fallen from 88 per cent of Britain's labour force in 1688 to 50 per cent before 1780; to 25 per cent by 1840; and to 2.2 per cent by 1988. Per capita productivity increased by a factor of 70:1 in 200 years.

The creation of factories followed, and the harsh working conditions led, after a bitter struggle, to the creation of trade unions (then called 'combinations'), and many trade unionists were transported to Australia.

The Industrial Revolution

The term 'Industrial Revolution' was first used by Auguste Blanqui in 1837, taken up by Engels in 1845 and by Marx in 1863, but adopted in English surprisingly late, in 1881.

Historians disagree on the precise dating of the Industrial Revolution in England, but it began somewhere between 1760 and 1780, was then extended to Scotland, the future nation of Belgium, and then to North America, Germany, France, and Italy.

Its essential feature was the transition from water and animal power to the use of large-scale steam power, dependent on coal, water, and iron. James Watt and Matthew Bolton designed and manufactured the double-action steam engine (1782), followed by the first steamboats (1783) and the first steam-driven textile factory (1785).

Precision engineering was essential. Iron replaced wood, or concrete, in bridges because of its strength relative to weight, and Thomas Telford's structures are still in use. There was heavy investment in canals, and then railways.

With the division of labour, a process such as making pins (to quote Adam Smith's famous example) involved having workers devoted to a single task that was then co-ordinated, bringing the shank and the head together. In North America, Eli Whitney pioneered the manufacture of standardised interchangeable parts, so that a rifle was not a unique product of a craftsman but part of a mass production process.

Friedrich Engels argued that the clock, even more than the steam engine, was the central tool of the Industrial Revolution, enabling co-ordination for systems (especially railways in the nineteenth century) and imposing uniformity in running schools, factories, mines, and shops.

There were major victims of the Industrial Revolution, which relied on sweated labour, especially in mines; the exploitation of women and small children; and the adoption of the principle of *laissez faire*, with governments refusing to intervene to improve working conditions and overcrowding in towns, which had inadequate water supply, no sewerage, and high levels of infectious disease.

The First Fleet arrived in Australia in 1788, during a decade of extraordinary change and accomplishment in Europe and North America, culminating in George Washington's inauguration as president of the United States (1789) and the outbreak of the French Revolution (1789). As mentioned earlier, Joseph II tried to organise the Holy Roman Empire on rational, utilitarian lines.

The Montgolfier brothers developed hot-air balloons (1783), followed by the first manned international flight, from France to England (1785).

The elements tungsten (1781), tellurium (1782), uranium (1789), and zirconium (1789) were discovered. The composition of water was determined by Cavendish and Lavoisier (1784), and Lavoisier also identified respiration with combustion, wrote the first modern chemistry textbook, developed the first systematic nomenclature, and proposed the law of the conservation of mass (1783–89). James Hutton published his 'uniformitarian' theories in geology (1788).

Mozart's greatest operas, symphonies, concertos, and chamber music were composed (1782–91), Haydn was exceptionally productive, and the young Beethoven was rising.

The fundamentalist revival

Contemporary attacks on Western liberal values shaped by the European Enlightenment come from divergent areas, but they have a striking and unexpected convergence.

Muhammad expressly promoted the toleration of Christianity and Judaism in the *Qur'an* (although he was very harsh on apostasy), a position that mainstream Muslims maintain. However, a radical fundamentalist minority, including jihadist supporters of the Islamic State, the Taliban, al-Shebaab, and al-Qaeda, loathe many Western institutions and use terror as a weapon. (Hezbollah uses violence but is primarily secular, not at all jihadist, although to victims the distinction may not be of much consolation.)

The characteristics and beliefs of Islamic fundamentalism include the following:

- an insistence on rule by theocracy and the inerrancy of scripture;
- dogmatism and inability to comprehend differing points of view;
- hostility to science and scientific method;
- belief in miracles;
- belief that the creation of the Earth was relatively recent, humans have not changed since their creation, and the theory of evolution is wrong;
- rejection of evidence-based decision-making;
- hostility to modernity;

- resistance to feminism and the changing roles of women;
- hierarchy and paternalism;
- homophobia;
- self-definition as 'fighters', and cultivation of a siege mentality;
- constant need to have enemies, to provide the rationale for forceful action;
- dismissal of United Nations and international opinion, including disregard for UNESCO World Heritage sites;
- suppression of dissent, including punishing critics or whistleblowers;
- readiness to act outside the rule of law, and to assume that 'the end justifies the means';
- assertion of simple explanations for complex problems; and
- deep suspicion of 'the Other'.

Disturbingly, many, perhaps most, of these beliefs and characteristics are shared to some degree by Christian fundamentalists, who comprise a significant proportion of the hard right in the United States, in some European states, and even in Australia. The Israeli government also includes some strong fundamentalist elements.

Jesus has been essentially hijacked by hard-right religious fundamentalists who seem to be unfamiliar with the Beatitudes or the parable of the Good Samaritan, and do not recognise him as having been a child refugee, sympathetic to women, non-judgemental, and a victim of judicial abuse and capital punishment.

Paradoxically, religion is exerting a stronger political influence in the twenty-first century than it did in the twentieth, despite the fact that, in the West, religious adherence has dropped, and churches have come under unprecedented, and long overdue, censure for their role in covering up the sexual abuse of children and their authoritarian treatment of Australia's Indigenous children.

Curiously, blasphemy is still a crime in New South Wales,

although there has been no prosecution of it since 1871.

The Ayatollah Khomeini and Margaret Thatcher shared four fundamentalist elements: a conviction of their infallibility, scepticism about 'progress', a commitment to absolutes, and invoking a Manichean contest between good and evil.

Enlightenment now

Steven Pinker, a Canadian-American cognitive psychologist and prolific author who holds a chair at Harvard, wrote the bestseller *Enlightenment Now: the case for reason, science, humanism and progress* (2018), which Bill Gates hailed as 'my new favorite book of all time'.

Pinker describes the Enlightenment 'as a cornucopia of ideas, some of them contradictory, but four themes tie them together: reason, science, humanism, and progress', adopting a definition by the physicist David Deutsch that 'optimism ... is the theory that all failures—all evils—are due to insufficient knowledge ... An optimistic civilization is open and not afraid to innovate, and is based on traditions of criticism.'

Pinker argues passionately that by adopting Enlightenment methodology we can overcome the major problems of mankind: we can live longer, healthier lives; dramatically reduce poverty and destitution; improve access to education; and give better opportunities for women and girls. He is sharply critical of 'morose cultural pessimists', and hostile to 'intellectuals', who ought to be his natural allies. The work is encyclopaedic in range, packed with information: one critic called it a 'Gradgrindian juggernaut' (named for Dickens' character Mr Gradgrind, who was an obsessive fact collector).

He concedes that the Enlightenment thinkers failed to grasp the concepts of entropy, evolution, and information theory, all of which were developed in the nineteenth century.

Since the end of World War II, from 1945 to 2020, there has been an unprecedented improvement in the physical quality of life. There have been no major wars (with the exception of Korea and Vietnam), although the Middle East, much of Africa, and Cambodia have been racked by bloody localised conflicts, and there has been considerable success in tackling disease, notably in the elimination of smallpox and the reduced incidence (so far) of measles, polio, and AIDS. Malaria, and now the coronavirus, remain a huge challenge.

But these great changes, including the ending of the Cold War and removing the fear of nuclear annihilation, depended on high levels of co-operation and collaboration, on trust between leaders, on seeking common ground, and on working with international institutions, especially the United Nations and its agencies, the World Health Organization, the Food and Agriculture Organization and the United Nations Educational Scientific and Cultural Organization.

Pinker is not particularly concerned about rising levels of inequality, preferring to emphasise the fact that abject poverty and malnutrition are reducing, despite rapid population increases.

He is predictably anti-Trump, and worries about excessive polarisation in politics, but this does not challenge his essential optimism. For example, he is deeply committed to the abolition of the death penalty worldwide—an Enlightenment goal that is promoted by the United Nations. Indeed, it is a condition of membership of the European Union that nation states not have the death penalty in their statutes.

The United States is the only Western democracy that still executes prisoners, admittedly in a minority of states (all of them in Trump territory). Pinker draws consolation from a chart indicating that the number of executions per 100,000 people in the United States had fallen from 0.85 in 1780 to 0.03 in 2015.

On climate change, Pinker sees the adoption of nuclear power as the best option, but acknowledges the psychological resistance to it, and the huge costs involved in building plants, providing for safety and environmental security, and—ultimately—decommissioning. He explains the science behind global warming very well, but his chapter on 'The Environment' is perhaps the weakest in the book, marred by a glib optimism.

He identifies the enemies of the Enlightenment as religion, theistic morality, authoritarian populism, and nationalism and tribalism, compounded by Romanticism.

He has a fixed conviction, almost an obsession, that intellectuals, especially in the humanities, are deeply opposed to science and to the scientific method, asserting that many of them see it—in the spirit of post-modernism—as just another cultural narrative, shaped by gender studies ('Glaciers, Gender, and Science') and sociology. From my own observation and experience, I reject this categorically.

What can we learn from the Enlightenment?

The fundamental lesson to be learned from the Enlightenment is that, in problem-solving, we must start with curiosity, examine the phenomena around us, apply rational analysis to them, form a set of testable propositions, disseminate them widely, and secure both public understanding and the willingness of leaders to apply the science.

Climate change is the greatest problem of our times, but our leaders are incapable of dealing with it. Climate change creates existential challenges to an increasingly fragile liberal democracy, comprising:

- the increasingly rapid destruction of habitat and the extinction of species;

- the destruction of ecosystems, creating famine and mass exoduses;
- an international refugee crisis, caused in part by rising sea levels and massively depleted environments, leading to the harsh treatment of asylum seekers and the radicalisation of the victims; and
- the abandonment of evidence-based policies to preference very short-term personal advantage (as in favouring lower electricity bills v. saving the planet).

If we start with the politics—especially politics shaped by the lobbying of vested interests—we will always get it wrong.

Contemporary issues such as censorship, surveillance, a free press, slavery, racism, arbitrary rule, church-state relations, religious tolerance, torture, the death penalty, and the separation of powers were all central to the Enlightenment project. And there are continuing, timeless issues from that era as well—such as the search for truth, understanding nature, and asking what education is for and what humans are capable of.

As citizens, we can learn a great deal by informing ourselves about the Enlightenment. We are still its heirs.

CHAPTER FOUR

How the Digital World Changed Everything

A popular Government, without popular information, or the
means of acquiring it, is but a Prologue to a Farce or a Tragedy;
or perhaps both. Knowledge will forever govern ignorance:
And a people who mean to be their own Governors must arm
themselves with the power which knowledge gives.

– JAMES MADISON (1822)

In March 2020 there were estimated to be 4.6 billion users of
the Internet, 59 per cent of the world's population. In the third
millennium of the Common Era (CE), as we now call it—since 'AD'
has no cultural resonance in China, India, Indonesia, Pakistan, or
Japan—digital technology gives people, for the first time in history,
global access to an almost infinite information resource, for free,
and communication access at relatively low cost.

Three men, all born in 1955 by a neat coincidence, played a
major role in transforming the modern world: Bill Gates, Steve
Jobs, and Tim Berners-Lee. Gates founded Microsoft, amassed
a fortune of $US102 billion, and became a great philanthropist;
Jobs, the driving force at Apple, left $US10.2 billion; while
Berners-Lee, the architect of the World Wide Web, is worth just
$US10 million.

After World War II, there was a hierarchical model of computing. The big computer—generally a 'special purpose' machine, huge in size, cost, and power consumption, but relatively limited in output—was a metaphor for centralised control of information processing, costing millions to build, and millions to operate and maintain. It was the preserve of governments—especially the armed forces and security services—universities, and corporations. Those huge early computers ran on expertise, providing input and interpreting output, and used more power than a locomotive. Now they can fit into a pocket or purse.

The development of microchips and the practice of miniaturisation democratised computing. Miniaturisation was the decisive technological achievement of the twentieth century, marking a radical discontinuity with past economic history. Two functions could be maximised simultaneously, enabling greater capacity at lower cost and extraordinary flexibility, so that we could—if so minded—buy shares or bet on a horse race while climbing Mt Everest.

Today, the processing power of the smartphone is 100,000 times greater than the guidance computer used on Apollo 11 for the Moon landing in 1969. Access to the Internet with a handheld smartphone enabled the revolutionary transformation of the World Wide Web.

This has profound implications for democratic practice. If every citizen has the potential to master the universe through a device in her/his hand, we might reasonably expect an unprecedented level of informed public discourse and a determination to solve the great challenges to civilisation, our habitat, and the planet itself.

It has not worked out that way.

An unexpected result of providing low-cost access to the big, wide world has been the rise of populism and identity politics, in which expertise is dismissed or derided.

Far from a heralding a new Renaissance or Age of Enlightenment, the almost universal ownership of computing devices and free access to the Internet and the World Wide Web has not led to the hoped-for expansion of knowledge, exposure to new experiences, understanding of our common humanity, and determination to provide rational solutions to complex problems, including racism and poverty, ending state violence, and wrecking natural resources.

Instead, there has been a downside, in which the digital revolution provides misinformation on an epic scale, and consumers' cherry-picking of sources results in 'confirmation bias', which reinforces existing prejudices. The development of politics as a spectator sport, or television reality show, created opportunities for the rise of leaders such as Silvio Berlusconi, Donald Trump, Boris Johnson, Rodrigo Duterte, and Jair Bolsonaro, and expanded the power of Narendra Modi, Recep Tayyip Erdoğan and Scott Morrison. Social media has provided unprecedented diversity, but without mechanisms for checking and testing.

Religious fundamentalism has been revived, and the cult of the individual has displaced a sense of community and the public interest. There have been sharp increases in gambling, drug use, conspicuous consumption, obesity, racist attacks, nationalism, suspicion of 'the Other', and a distinct anti-science bias.

If anything, recent decades have been marked by the fragmentation of knowledge, linked to individual experience and short-term needs. Now every user can be her/his own expert and seek the variety of truth that best suits them.

The Internet (1988)

The Internet, originally named ARPANET (Advanced Research Projects Agency Network) and part of the Department of Defense, was proposed by the Rand Corporation, a global policy think tank, and launched in the United States in 1969. Originally a network of research computers in American universities, it was designed to share knowledge with selected users during the Cold War, and to protect findings, even after nuclear attack, by dispersing them.

The name 'Internet' was used informally between 1970 and 1975, and adopted formally in 1982.

In 1988, the Internet was handed over to the National Science Foundation in Washington, and rapidly became a 'network of networks'—government, academic, commercial, and private—operating worldwide. Users themselves developed the Internet, which was voluntary and open, described by former CEO of Google, Eric Schmidt, as 'the largest functioning anarchy in the world'. It illustrated Marshall McLuhan's concept of 'the global village' and refuted the adage that 'There's no such thing as a free lunch.'

The then US senator Al Gore secured passage of *The High Performance Computing Act* (1991), known as 'the Gore Bill', which promoted the Internet as 'an information super-highway'. (His claim to have 'invented the Internet', intended as a joke, was used against him, showing how dangerous humour can be in politics. But Gore's role with the Internet was significant.)

In 1990, there were 313,000 users of the Internet; 10 million by 1996; one billion by 2005; and over four-and-a-half billion by 2020.

The World Wide Web (1989)

The year 1989 proved to be one of those turning-point years in history—highlighted by the demolition of the Berlin Wall, the end of the Cold War, the Tiananmen Square massacre, Nelson Mandela's release from prison, growing recognition of the existence of a post-industrial economy, the first battery-powered personal computer, and the creation of the World Wide Web (WWW).

The WWW is the software that links 'hypertext' documents to form a 'web', using the Internet as the carrier.

The impact of the web has been compared to Johannes

Gutenberg's development of the printing press with moveable type, in about 1450, which transformed the culture, religion, economics, and politics of Europe—and then spread throughout the world. But the web had even greater impact, spreading far more rapidly and with unprecedented penetration.

Tim Berners-Lee, an English physicist and computer scientist, designed the web in 1989 while working as a consultant for CERN (the European Organization for Nuclear Research) in Geneva. He developed the necessary tools, including a web server and a web browser, the Hypertext Transfer Protocol (http://), and Hypertext Markup Language (html).[1]

The web was made available to the general public in August 1991, and Berners-Lee campaigned to ensure that it remained 'open, non-proprietary, and free'.

Berners-Lee retained a childhood memory of a late-Victorian reference book, *Enquire Within Upon Everything*, that provided advice on a vast array of subjects. It went through 110 editions between 1856 and 1916. At one level, the contents could look random, but it was efficiently linked and very well indexed. As a child, I had a copy myself, but failed to take it up as Berners-Lee did.

Nowadays, if you want to access material from the Louvre or the Vatican Library, or to listen to a Beethoven symphony, it is only a fingertip away if you have access to Google, YouTube, or *Wikipedia*. However, if you are pursuing child pornography, catching up with a Nazi group, learning how to blow up a mosque, or finding the nearest drug dealer, Google and other websites can help you just as well.

But every search leaves a trail. So if you keep accessing entries on explosive devices, or gambling, or gang rapes, or motorbikes, or Thai sex tours, some other searcher, unknown to you, will, before long, be able to assess your needs, tastes, and propensities, and sell them to a third (or fourth) party. And you may become a person of interest to security agencies.

What we seek defines who we are.

In late 2019, I checked details about the history of Copenhagen and the musical tradition of Dresden on *Wikipedia*. Soon, I began receiving unsought advice from travel agents about flights between Copenhagen and Dresden, and accommodation offers. This can be helpful—and disturbing. Every enquiry is a saleable property. Someone (or an advertising-linked algorithm) is always watching. There are no secrets.

If I want to re-read my own speeches from years ago, it is far quicker to use the web than to explore my own files.

Apart from its obvious benefits, the web is ideal for communicating propaganda and targeting the vulnerable or susceptible. Information retrieved from patterns of social-media use has been used to interfere in the conduct of elections and referenda. It is also a perfect tool for marketing, because consumption patterns revealed from the Web can then be sold on. This is central to the operation of social-media platforms.

In his 2018 Leslie Stephen Lecture in Cambridge, 'Liberalism, Populism and the Fate of the World', the historian Simon Schama argued, 'The World Wide Web turns out not to further world wide truth and transparency but instead has been a perfect nesting place for communities of self-reinforcing myths, lies and conspiratorial fantasies.'

Sleepers, Wake! made the same point in its 4th edition (1995) on p. 184:

Phillip Adams argues that the real threat of the communications revolution will not be the homogenisation of cultures but 'peoples' desire to be tribal rather than truly national, let alone global ... [T]hese are not mutually exclusive concepts ...

[M]embers of local tribes will be able to communicate instantly and comprehensively with like-minded people—members of the same tribe—all over the world'.

And this may include those who are obsessed with pornography, anti-Semitism, or weaponry.

The smartphone (1995–2007)

The first handheld mobile phone, developed by Motorola in 1973, weighed 2 kilograms and could be used as a blunt instrument. The size, weight, and capacity of mobile phones changed dramatically in the next decade, and they became generally available from 1983.

The term 'smartphone' dates from 1995; Apple's iPhone was launched in 2007.

The smartphone, in addition to its original function as a mobile telephone, now includes a camera for single shots and film; a two-way interactive video and sound recorder; photo storage; a clock, an alarm, and a calculator; a keyboard; email, texting, and Twitter resources; and access to radio, music, newspapers, journals, Wikipedia, weather reports and forecasts, web browsing, maps, satellite navigation, games and gambling, banking, buying goods, and ordering—and tracking—food and transport.

Contrary to what might have been expected, far from exploring the universal and the long term, the new digital technologies have reinforced the realm of the personal, as demonstrated by social media, with its emphasis on the immediate and the tribal, on family and close friends, and on reinforcing existing views rather than enabling challenging exposure to new ones.

Smartphones and the internet have changed behaviour, time use, and employment, and have challenged conventional education more deeply and rapidly than any other technology—even the motor car, radio, film, or television.

The smartphone has become the new best friend, a love object, the last thing seen and touched at night, the first thing seen and touched in the morning. For many, life experience seems to be as an appendage to the device.

Direct observation of fellow citizens confirms this, every day, as they cross busy roads, eyes fixed on the device. A combination of the smartphone and dark glasses makes it easy to avoid eye contact with other humans. There is little sense of community: subjectivity is all. 'Feeling' is central. 'Knowledge' is optional.

This weakens our sense of, or empathy with, 'the Other', the remote, the unfamiliar, and devastates our sense of being members of a broader community. Now, individual need is not just the *primary* motivator, but the only one.

The average time spent on a smartphone each day is estimated at four hours, with some addicts regularly spending nine hours online, including on smartphones. Being without one's smartphone may create a sense of isolation and panic—what some psychologists have called 'nomophobia'.

Did anyone predict the addictive, even obsessive, use of the smartphone? I can find no material that answers this question, so I assume not. I certainly failed to do so.

I once thought about writing a short story—a horror story, really—about a society in which everybody between the ages of 15 and 30 was required, when in public, to be connected to a handheld electronic device and to have it in view at all times, subject to heavy penalties, as part of a surveillance state, so that the subject is constantly monitored, and serious face-to-face conversation, let alone serious reading or serious thinking, becomes impossible. Too draconian? Impossible? Readers would judge it to be too far-fetched. Surely it could never happen?

Social media (1995)

There are claimed to be 3.8 billion active users worldwide of social media, where Internet platforms enable people to connect with others on the basis of a common interest.

Its history is rather vague. The essential pre-conditions for social media were email (electronic mail) and the Internet. Email was the precursor, and remains central.

Email was invented by the American computer programmer Ray Tomlinson. It first operated on the ARPANET system in 1971, becoming universal after development of the Internet. It transformed how billions of people communicated—no longer on paper, providing instant connection and rapid response. Historically, letters required careful thought before dispatch, and had a single recipient. Now, emails, often containing sensitive or damaging material, can be sent throughout the known world to multiple recipients, and cannot be retrieved. As Timothy Snyder wrote, 'Remember that email is skywriting.' The turnaround time is not days, but minutes, even seconds.

The earliest example of social media—although not being used with that name—was in 1979 with USENET, which exchanged messages between students at Duke University and the University of North Carolina.

The term 'social media' first appeared in the 1990s, and the first network, SixDegrees.com, operated between 1997 and 2001, followed by Friendster.com, LinkedIn.com (for professional networking), and MySpace (e-commerce, but linking high-profile performers with their fans).

Later, the web transformed and expanded social media beyond recognition, and its economic potential grew exponentially. So did its social and political implications.

Users can now choose their own evidence by using the web as the gateway to the Internet, through social-media platforms and search engines such as Facebook, YouTube, WhatsApp, Instagram, Twitter, Google, LinkedIn, and Skype, and, in China, WeChat, to list only the most used.

Facebook: privacy, hacking, stalking, and trolling

Facebook, the dominant force in social media, was founded in 2004 by Mark Zuckerberg (1984–) and some fellow Harvard undergraduates. Originally targeted at college campuses, it was launched publicly in 2006. Zuckerberg dropped out of Harvard and relocated to Palo Alto, Silicon Valley, in California.

In 2012, when Facebook became a public company, it already had one billion users. Today, Facebook has more active followers globally (2.6 billion) than there are notional Christians (2.3 billion). As Evan Osnos notes: 'If Facebook were a country, it would have the largest population on earth.'[2]

Facebook also owns Instagram, WhatsApp, Oculus, and Facebook Messenger. Advertising accounts for 98 per cent of its revenue, greater than every US newspaper combined. On 31 August 2020, its market capitalisation was $US835 billion.

Facebook and other social-media sites also depend on content generated by the user. The user produces the raw material for social media as a free gift, which the sites can then dissect, aggregate, and sell off to third parties. The consumer is the subject-matter provider. There is no inhibition about invasion of privacy: Facebook users insist on it.

If a black car with tinted windows were to turn up at a citizen's front door, and its occupants took away her/his private papers and diaries, it would be regarded as an outrageous violation of civil liberties, because something physical had been taken. With a social-media site, the same material in digital form is given up freely and openly for general consumption.

In 2016, during the United Kingdom's referendum on leaving the European Union, and the US presidential election, hackers used Facebook to boost the 'Leave' and Trump campaigns, in order to break up a united Europe and weaken, if not eliminate, US support for multinational organisations and arrangements such as the United Nations, NATO, and the Paris Accords on climate change.

The hacking was directed from Russia through Facebook. 'Fake news' was disseminated from 100 websites in Veles, Macedonia, with advertisements directed at 'micro-targets'—people marked by 'isolation, outrage and addictive behaviour'. The Internet Research Agency, an under-the-radar instrument of Vladimir Putin's government, created Facebook groups, aiming at gun-rights enthusiasts and generating anxiety about secure borders and Hillary Clinton's integrity. Their Facebook campaigns were estimated to have reached 150 million Americans.

Facebook allowed the political-consulting firm Cambridge Analytica to harvest information from about 50 million Facebook users—later increased to 87 million—enabling the development of psychological profiles of potential voters.[3]

Further abroad, lynchings in India and the persecution of Muslims in Myanmar were directly linked to the use of Facebook.

In the US, in 2019, the Trump re-election campaign produced more than 218,000 different Facebook advertisements, aimed at reinforcing support from conservative potential voters. Trump's social-media guru, Brad Parscale, said, 'The campaign is all about data collection ... we want to know who you are and how you think ...' The advertisements were seen more than 600 million times, and the most common theme was attacking any form of criticism of Trump as illegitimate, and dubbing it 'fake news'.

In the 2019 Australian election, Liberal Party Facebook videos were viewed 17,630,800 times, compared to 5,940,500 for the ALP, and the advantage increased to 4:1 in the last stages of the campaign. According to Christopher Knaus in *Guardian Australia* (4 June 2019), 'No Labor video received more than 500,000 views, while the Liberals had seven [that did].'

For the first time, in 2019 the parties advertised more on social media and digital platforms than in newspapers, and on television and radio.

Social media and other tools of the digital revolution have been important aids in recruiting some users to extremist causes, and for 'trolling'. If you have a commitment to freedom of expression, as Zuckerberg claims, should Facebook provide a platform for Holocaust deniers? Until late June 2020, Zuckerberg had ruled that there would be no fact checks for campaign material published on Facebook.

Trolling, the act of using social media, anonymously, to pursue an individual—perhaps a child at school—by posting inflammatory messages on an online news group, chat room, or blog—has had devastating effects on people's confidence and mental health, often leading to social withdrawal, drug dependence, or even suicide.

A high proportion of mass shootings have been committed by perpetrators who set out their grievances against the world, and particular groups in it, often at great length, on Facebook and other social-media sites. Presumably, Facebook users become inured to constant threats of revenge.

The controversial #MeToo campaign exposed many appalling cases of sexual harassment. However, it also led to the particularly virulent trolling of vulnerable women who were subject to vicious attacks that could then be transmitted throughout the world. The Internet is also particularly valuable for stalkers.

It can take years to build up a reputation—or a sense of security—and only seconds to destroy it. The target is damaged while the perpetrator is unscathed.

Social media has given political actors the capacity to reach more people more of the time in more places than at any stage of human history. But are the messages accurate? Are they based on evidence? How can they be tested?

Twitter, a microblogging and social-networking service, was founded in 2006, and has more than 320 million active users. Its messages ('tweets') were originally limited to 140 characters, doubled to 280 in 2017. Jack Dorsey, the co-founder of Twitter, admits that he never contemplated its negative aspects.

President Trump became the world's most famous tweeter (@realDonaldTrump), sometimes sending more than 100 tweets in one day, clearly all his own work. It is a perfect medium for him, with no scope for nuance or analysis, but good for snap judgements or bullying. One can respond to a tweet, but there is no scope for argument.

Trump was enraged by Twitter's proposals to add labels and context to inflammatory tweets and to apply fact-checks, insisting that the Constitution enshrined the right to lie.

The 'Big Five' technology giants + Wikipedia

The Saudi Arabian oil and gas behemoth Aramco (a.k.a. Saudi Aramco) used to have the world's largest market capitalisation: it was $US2.1 trillion in December 2019, before falling a little in 2020.

It has since been overtaken by Apple, which heads a list of five American IT-based companies—the others being Microsoft, Amazon, Google, and Facebook—with a total market capitalisation in excess of $US7.5 trillion, well ahead of 'Big Oil', 'Big Media', motor manufacturing, and aviation.

At the time of writing, Apple's market capitalisation was $US2.2 trillion; Amazon's, $US1.7 trillion; Microsoft's, $US1.7 trillion; Google's (Alphabet's), $US1.1 trillion; and Facebook's, $US835 billion.

All five reflect the technological dominance of the United States since the Second Industrial Revolution—the electric revolution—due to the work of Thomas Alva Edison, Alexander Graham Bell, Nicola Tesla, and George Westinghouse.[4]

All five touch and shape our lives. So does *Wikipedia*, although it has no significant market capitalisation.

Apple (1976)

Apple Inc., founded in Cupertino, Santa Clara Valley, California, in 1976 by Steve Jobs (1955–2011) and Steve Wozniak (1950–), is now the world's largest company by market capitalisation. The apple logo is a nod in the direction of Isaac Newton's Eureka insight about gravity: a Macintosh (strictly, McIntosh) is a Canadian apple.

Wozniak was the engineer; Jobs, a designer and marketer of genius, developed the iPhone, iPad, iTunes, iPod, the Pixar company, and the laser printer. The Apple Macintosh personal computer was launched in 1984, and was rebranded as 'Mac' in 1998. The first battery-powered personal computer (PC) was available from 1989.

Wozniak and Jobs had a spectacular quarrel: Wozniak left Apple, and at one stage thought seriously of relocating to Melbourne.

Apple went through a dip in the years 1991 to 1997, Jobs dropped out and then returned, and the company returned to a high-growth trajectory, even after his death.

Apple has secured an exceptional, quasi-religious, degree of brand loyalty, even after coming under serious attack for its labour practices. (Most of its iPhones are manufactured by Foxconn in Zhengzhou.)

Microsoft Corporation (1975)

Bill Gates was a child prodigy, born in Seattle. He dropped out of Harvard, and at nineteen began designing software for the Altair 8800, a stand-alone microcomputer. He developed the software packages 'MS-DOS' and 'Windows', and decided to establish his own software firm, virtually creating a new industry, which would contract out the manufacturing.

Bill Gates became the first software billionaire, and set up the Bill & Melinda Gates Foundation, the world's largest charitable orgnisation, devoted to global health, especially polio and malaria. His wealth was estimated at $US107 billion in 2019.

Microsoft's headquarters are at Redmond, Washington State, near Seattle. The company bought LinkedIn and Skype, and is heavily involved in cloud computing (on-demand data storage via the Internet).

Google (1998)

Google was founded in Silicon Valley, California, in 1998 by Larry Page (1973–) and Sergey Brin (1973–), two Stanford PhD students.

The name Google was a mis-spelling of the word 'googol', which in mathematics is the number 1 followed by 100 zeroes. This is not a number we are likely to use much in everyday life—it is probably more than all the atoms in the universe.

In 1996, Page and Brin devised PageRank, a 'ranking algorithm', and Google explains:

> PageRank works by counting the number and quality of links to a page to determine a rough estimate of how important the website is. The underlying assumption is that more important websites are likely to receive more links from other websites … A PageRank results from a mathematical algorithm based on the webgraph, created by all World Wide Web pages as nodes and hyperlinks as edges.

Google effectively destroyed the market value of all other search engines on the market with its ease of access to material all over the world.

Its motto is 'Don't be evil.' Google Search boasts the world's most-visited website. It has more than one billion users, and stores more than 2 trillion (2,000,000,000,000) files.

It developed services such as Google Chrome (a web browser), gmail (that is, Google mail), Google Maps, Instagram, 'icloud' electronic storage of data, the simulation of virtual reality, and hardware—the Nexus computer and Pixel smartphone. Google is now working on artificial intelligence (AI) and medical diagnostics. YouTube is a Google subsidiary. Google Earth presents 3D imagery of the planet, generated by satellite, capable of identifying houses in a street and individual trees in a forest.

Google is now part of the conglomerate Alphabet Inc, with Page as its CEO. Page and Brin are currently ranked no. 9 and no. 10 in the list of the world's richest people.

Amazon.com (1994)

Jeff Bezos (1964–), born in New Mexico, educated at Princeton, worked in Wall Street, then relocated to Seattle so that he could have easy access to Microsoft. In 1994, he founded Amazon.com in Washington State as an online retailer, selling books at first, and then rapidly expanding to electronics, games, toys, furniture, domestic white goods, and jewellery. By 2015, Amazon had greater market volume than the Walmart supermarkets. Bezos then led Amazon into publishing, film and television programs, video streaming, audiobooks, and music. Amazon also bought bricks-and-mortar stores that sold food, and developed Amazon Prime, a subscription delivery service. Amazon even ventured into aerospace manufacturing, and Blue Origin orbited into space in 2015

Bezos bought *The Washington Post* in 2013. In 2018, he was identified as the world's richest man, worth $US127 billion in 2020. He came under sustained Twitter attack from Donald Trump, who accused *The Washington Post* of publishing fake news. He was also criticised for tax avoidance. Having made only modest contributions to philanthropy, in February 2020 he created the Bezos Earth Fund, with capital of $10 billion, to fight climate change.

Facebook (2004)

Mark Zuckerberg (1984–), the founder, remains the dominant force in Facebook (see p 96). With his wealth estimated at $US62 billion, he ranks no. 8 among the world's richest people, and is by far the youngest of them.

Wikipedia (2001)

Wikipedia has become an essential tool in the digital revolution, but is run by a not-for-profit foundation. *Wikipedia* was principally created by Jimmy Wales (1966–), born in Alabama, but now living in London. (Larry Sanger insists he was co-founder.)

Originally based in San Diego, California, and now in San Francisco, *Wikipedia* went online in January 2001 and is available, free, in 305 languages. There are 6 million entries in the English-language *Wikipedia*, the equivalent of 49 million pages and 1.5 million biographies (many of them 'stubs'). The German and French versions both have over 2 million entries.

'Wiki' is the Hawaiian word for 'quick'.

The IT Revolution has been described as a 'digital hurricane', and reference books were early victims. *Wikipedia* completely transformed reference publishing: few hardback or paperback reference works survived. *Nature* examined *Wikipedia*'s science entries, and concluded that generally they were as accurate as *Encyclopædia Britannica*. An Oxford University study in 2013 concluded that the most contested entries had a religious connotation—for example, Muhammad, Jesus, Christianity, abortion, and circumcision. While the opportunity for fake and distorted entries exists, so does the opportunity for immediate challenge and correction.

Longer articles, such as 'Periodic table' or 'J.S. Bach', are often outstanding—but more than half are of comparatively little, or no, general interest ('Iona Gaels Men's Soccer').

Jack Lynch, in his admirable *You Could Look It Up: the reference shelf from ancient Babylon to Wikipedia* (Bloomsbury, 2016), comments: '*Wikipedia*, despite being non-commercial, still poses many of the dangers of a traditional monopoly, and we run the risk of living in an information monoculture.'

Wikipedia's 'five pillars' are: it is 'an encyclopedia ... not an indiscriminate collection of information ... not a soapbox, an advertising platform, a vanity press, an experiment in anarchy or democracy, or a web directory'; it is written from 'a neutral point of view', which means it strives for 'articles that advocate no single point of view'; it is 'free content that anyone may change'; it has a code of conduct, requiring contributors to treat each other with respect and civility; and it 'does not have firm rules.'

When I check a reference or a quotation on *Wikipedia*, it can be disconcerting to find myself listed as the authority.

The pioneers

We have come to accept digital technology as a given, as if it has always been with us. This demonstrates the problem of short-term collective memory.

The background story is worth pursuing.

The abacus is the oldest digital calculating instrument, first used by Sumerians in Mesopotamia about 2700 BCE, much later in Persia and China.

Calculating rods were employed when the Scot John Napier produced his logarithmic tables (1617), and William Oughtred developed the slide rule (1621).

Blaise Pascal (1623–62), the prodigious French mathematician, philosopher, and theologian, author of the celebrated *Pensées*, devised, at the age of nineteen, the first digital calculating machine, which had cogged wheels and could add and subtract.

Automata were developed in France, Switzerland, and Italy as entertainments and in toys, such as music boxes. Jacques de Vaucanson (1709–82) achieved fame with his flute player, tambourine player, and 'The Digesting Duck', with defecation as a special feature.

Joseph-Marie Jacquard (1752–1834) invented a loom (1804) in which punched program cards were wrapped around a cylinder through which air was blown. This could weave silk threads automatically to a predetermined design, texture, and colour. After improvement by others, the Jacquard loom could produce extremely elaborate and beautiful representations of people, animals, and plants.

As early as 1823, Charles Babbage (1791–1871) had proposed a machine for tabulating mathematical calculations for up to 20 decimal places. With government support, he worked on the first analytical computer, the programmed mechanical 'difference engine', but it was left incomplete (1842). He then designed 'Difference Engine No. 2' (1846–49), but it was not built until 1991 and is now on display in the London Science Museum.

From 1837, Babbage planned an 'analytical engine' to be fed by sets of Jacquard punch cards read by mechanical 'feelers', but it was never finished. He worked with the remarkable Ada Lovelace (1815–52), daughter of the poet Lord Byron, and by marriage, Countess of Lovelace. If not quite the world's first programmer, she published the first algorithm for a computer (1843), and, also influenced by Jacquard's automated textiles, went further than Babbage in speculating about what could be achieved with symbols, not just numbers, in a general-purpose computer. She was interested, she wrote, in 'poetical science'.

Bell Labs (1880)

Alexander Graham Bell, the inventor of the telephone, established a laboratory in Washington in 1880, modestly named after Alessandro Volta. It expanded, adopted Bell's name, was bought by AT&T, moved to New Jersey, and in 2016 was sold to Nokia and renamed as Nokia Bell Labs.

Researchers in the Bell Labs had an extraordinary record of discovery in radio, television, transistors, fibre optics, radio astronomy, lasers, photovoltaic cells, and electronic surgical tools. Fourteen of them won or shared Nobel Prizes.

IBM Corporation (1911)

IBM (originally International Business Machines), often referred to as 'Big Blue', was founded in New York in 1911. It became the market leader in office equipment, business systems, and large projects such as tabulating census data (which the Nazis found very useful in rounding up Holocaust victims in Europe).

From 1945 IBM invested heavily in basic research, and six of its scientists received Nobel Prizes. IBM employed lateral thinkers such as the mathematician Benoît Mandelbrot, the theoretician of chaos theory and fractals. IBM developed dynamic random-access memory, bar coding, magnetic-stripe cards, automatic teller machines (ATMs), and the floppy disk, and took out more patents than any other company. There are IBM research laboratories in Zurich, Melbourne, Dublin, Cambridge, Beijing, and Tokyo.

Boolean logic and the transition from analogue to digital

Modern computers have two common elements: they use binary logic, first proposed by George Boole, and they are digital, not analogue.

Victor Shestakov in Moscow (1935, but not published until 1941) and Claude Shannon at the Massachusetts Institute of Technology

(1937) independently recognised Boolean logic as being central to the operation of a universal computer.[5]

Boole (1815–64), an English mathematician and logician, working in Ireland, combined algebra and symbolic logic instead of the conventional methods for calculation. 'Boolean logic' uses binary numbers (pioneered by Gottfried von Leibniz, Isaac Newton's German contemporary and rival), in which quantities can be expressed by using only two symbols (1 and 0), which stand, in effect, for 'true' and 'false', or 'off' and 'on', as in the game of 'Twenty Questions', where each 'Yes' or 'No' limits the range of future questions. This led to the creation of NOT, AND, and OR 'gates' in computers, where contingent decisions result after one set of computations leads to another.

We live in an analogue world in which signals, as in real life, are transmitted continuously (represented by a sine wave in a graph), analogous to physical inputs, such as hearing a voice or watching a film. A traditional clock, with hands sweeping across the face, is analogue. However, analogue signals are subject to distortion, loss of quality over distance, and interference by noise. And analogue hardware is inflexible.

Computers live in a digital (based on 'digits', that is, numbers) world in which the signals (square waves on a graph) are discrete or discontinuous, expressed in binary numbers. The advantage of digital signals is that they are not affected by noise or distortion, distance is irrelevant, and digital hardware is very flexible. A digital clock displays numerals.

Blood-pressure monitors used to be analogue, with a needle on a calibrated dial. Now they are exclusively digital.

In his stimulating book *The Innovators: how a group of hackers, geniuses, and geeks created the digital revolution* (2014), Walter Isaacson drew attention to a striking convergence in the year 1937 of research on information processing.

Exactly a century after Babbage began planning his 'analytical engine', seven young men made major contributions to the central issues of information processing: Alonzo Church in Princeton; Alan Turing in Cambridge and Princeton; Claude Shannon at MIT; George Stibitz at the Bell Labs; Howard Aiken in Harvard; Konrad Zuse in Berlin; and John Vincent Atanasoff in Iowa.

Church paralleled Turing's work on the operation of a universal computer, while Shannon worked on Boole's binary logic and developed faster switching devices. Aiken designed the Harvard Mark I electro-mechanical computer (constructed by IBM engineers). Stibitz and Zuse improved complex relay systems borrowed from telephony. Atanasoff built a part-mechanical, part-electronic digital computer.

The most significant of the seven, Alan Turing (1912–54), a Fellow of King's College, Cambridge, studied with Church at Princeton, but had already published a seminal paper, 'On Computable Numbers, with an Application to the *Entscheidungsproblem*' (1937), referring to a challenge about the 'halting problem' made by the German mathematician David Hilbert. This contained a formal theory of universal computer operations, describing how encoded storage materials could be shuttled backwards and forwards, reading, calculating, and correcting as programmed. He invented an abstract model (often called a 'universal Turing machine') that embodied the logic of any computer, present *and* future.[6]

The early computers were expensive, cumbersome, costly to run, and prone to break down. Britain's Colossus, the world's first programmable electronic computer, developed in 1943–45 during World War II, was designed by Tommy Flowers of the General Post Office as a code-breaking device. It used 1,600 thermionic valves (vacuum tubes), and was located at Bletchley Park, Buckinghamshire. Ultimately, ten models were built. Colossus was not a general-purpose computer.

The operation of Colossus was a top secret, and when the war ended Winston Churchill ordered that it be dismantled and the documentation destroyed. Colossus remained a secret until 1974.

The first fully electronic computer, ENIAC (Electronic Numerical Integrator and Computer), designed by J. Presper Eckert, Jr, John W. Mauchly, and J.G. Brainerd, was built at the University of Pennsylvania (1946). It used analogue notation, weighed 30 tonnes, occupied a space of 90 cubic metres, and had 18,000 vacuum tubes/valves. It was a general-purpose computer and could be described as 'Turing complete'. A thousand times faster than electronic-magnetic-machines, it was first used to predict the feasibility of a hydrogen bomb.

By way of comparison, the modern smartphone has 1,300 times the capacity of ENIAC.

EDVAC (Electronic Discrete Variable Automatic Computer), built for the US Army Ballistics Research Laboratory and located in Maryland, was operational from 1947. Also designed by Eckert and Mauchly, it was the first successful general-purpose computer, with digital notation and stored internal programming. It had 6,000 vacuum tubes, weighed a sprightly eight tonnes, and required 30 people to operate it. Atanasoff claimed that some of his ideas had been plagiarised in its design, and a judge found in his favour.

EDVAC had critical input from the brilliant Hungarian polymath John von Neumann (1903–57), whose software programming included 'random' factors to approximate human decision-making. The architecture of modern computers is often ascribed to von Neumann, but he had been strongly influenced by Turing's insights.

Australia had the world's fifth stored memory computer, CSIRAC, built for the CSIRO (Commonwealth Scientific and Industrial Research Organisation), designed by Maston Beard and Trevor Pearcey, and completed in 1949. The input was on punched paper tapes, had 2,000 valves, and used 30 kilowatts of power. CSIRAC was kept secret, but became fully operational from 1951 to 1964 and is now on display at Scienceworks in Spotswood.

CSIRAC never engaged the interest of the Menzies government, which also failed to fund semiconductor research in Australia.

Miniaturisation: chips and microchips

In 1894, the Indian physicist J.C. Bose first used the interaction of metal ('the cat's whisker') and a crystal, mostly galena, to detect radio waves. For decades, enthusiasts—mostly young people—could listen to broadcasts on their crystal sets. The crystals were 'solid-state' semiconductors, using material such as germanium, silicon, or galena, in which an input signal disturbs relationships between atoms arranged in a 'lattice' formation.

Second-generation computers replaced vacuum tubes with small solid-state semi-conductors known as transistors. The transistor was invented in 1947–48 at the Bell Labs by William Shockley, John Bardeen, and Walter Brattain, who shared the 1956 Nobel Prize for Physics. Transistors performed the same function as the valve, but at a fraction of size, cost, energy use, or heat generation.

Then came microchips.

Jack Kilby of Texas Instruments developed a hybrid integrated circuit (embedded in germanium) in 1958, and received a Nobel Prize in 2000. Robert Noyce of Fairchild Semiconductor designed a microchip (monolithic integrated circuit chip), embedded in silicon, in 1959.

Their use revolutionised the cost, capacity, and size of the units.

The Anglo-American economist Kenneth Boulding argued (in 1964) that 'the scientific history of the Universe could largely be written in terms of three great concepts—equilibrium, entropy and evolution'. The second law of thermodynamics states that entropy always increases with time, 'entropy' being a 'running down' process in systems, a transition from order to disorder/randomness. Periods of equilibrium are succeeded by historic change governed by two apparently opposing processes: the 'running down' process associated with increasing entropy, and a counterbalancing 'building up' process associated with evolution.

Information has four properties that seem to challenge the 'running down' process: it is inconsumable: it does not disappear through use; on transfer to a recipient it remains with the sender; it is indivisible, usable as 'a set' rather than as a specific input such as water, food or electricity; and it is accumulative, added to, and reused.

In addition, the Internet and World Wide Web enable concentration, dispersion, circulation, and feedback.

Intel was established in 1968 by Robert Noyce and Gordon Moore in Santa Clara, California. (Its name is a contraction of 'integrated electronics'.)

'Moore's law' (1965) was a prediction by Gordon Moore that the number of transistors in an integrated circuit would double every two years, with reduced cost—leading to an exponential rise in capacity together with an exponential fall in total inputs, energy, labour, capital, space, and time.

As Tom Forester put it, 'If the airplane business had developed like the computer business ... a Boeing 767 would cost just $500 and ... circle the globe in 20 minutes on five gallons of gas' (*High Tech Society*, Oxford, 1987).

Intel specialised in microprocessor and chip design, and Federico Faggin led the team that produced the first micro-processor unit (MCU), a 'computer on a chip', in 1971.

A single-atom transistor, capable of opening and closing an electrical circuit, was demonstrated at Karlsruhe Institute of Technology in 2004, and has been perfected at the University of New South Wales by Professor Michelle Simmons.

Xerox invented the computer mouse and a graphical user interface (GUI), for working with images, but its desktop computer (1986) was heavy, awkward, and slow, and the company decided to concentrate on printers.

Mainframe computers, PCs, and super computers

The PC emerged in 1973, aimed originally at hobbyists. The first commercially successful PC was an Altair 8800, designed in New Mexico, using an Intel 8080 CPU (central processing unit) and Altair BASIC software designed by Bill Gates, and marketed in 1975.

The IBM System/360 was the dominant mainframe computer, both scientific and commercial, from 1965 to 1978, and the company diversified its production from microcomputers to supercomputers. A number of IBM 'clones' soon went on the market.

IBM's first PC, using Intel chips and Microsoft software, was marketed from 1981 to 1987. It soon found itself in direct competition with a rapidly rising Apple, which developed a graphical user interface and relied on an incompatible operating system.

After a spectacular revenue fall in 1993, IBM sold off much of its manufacturing and began diversifying into the provision of services such as consulting, commercial weather services, and cloud video, and the acquisition of property.

Seymour Cray (1925–96) was the pioneer of the supercomputer, culminating in the Y–MP 90 (1990). However, supercomputers were soon challenged by 'massively parallel' computing, whereby a network of smaller computers, often in their thousands, was linked synergistically to create greater capacity. Intel worked closely with IBM in developing parallel processing.

Supercomputers are still being built in the US, Japan, and China, and about 500 are in operation, most using Linux software devised by the Finnish-American engineer Linus Torvalds (named for Linus Pauling). Some supercomputers can perform over one hundred quadrillion (10^{15}, 10 to the power of 15) floating points per second (petaFLOPS). (A 'floating point' is where a decimal or binary point can move in manipulating huge numbers.)

Living in the digital world

In the Information Age, could daily life exist outside the ecosystem created by the 'Big Five' digital mega-corporations? Almost certainly not.

Their influence, both direct and indirect, is greater than we are aware of or care to admit, determining our patterns of expenditure, time use, addictions, and obsessions. They shape what we think we know, making privacy an outdated concept, redefining what it means to be an individual actor, with serious implications for surveillance and national security, and destroying trust in many institutions, the media, elections, and politics in general. Unexpectedly, unprecedented access to information has not reinforced the role of evidence and professional expertise; instead, it has come close to destroying them.

The digital world has revolutionised manufacturing, which operates globally. Typically, homes, offices, and retail outlets are stacked with electronic objects produced by companies they have never heard of, contracted out by firms with familiar names.

Among these are:

- Tencent, China (1998), incorporated in the Cayman Islands, headquartered in Shenzhen. Video games, gambling, social media, entertainment, e-commerce, real estate. Its largest investor is the South African investment giant Naspers Limited (founded in 1915). It runs WeChat, with more than one billion monthly users. Market capitalisation: $US649 billion. Employs 54,000 people.

- Samsung Electronics, South Korea (1969). The world's largest manufacturer of televisions, smartphones, chips, and lithium-ion batteries. Samsung, the parent company, founded in 1938, is the largest Korean *chaebol* (conglomerate), and includes shipbuilding and construction. Market capitalisation: $US326 billion. Employs 309,000 people.
- Foxconn, Taiwan (parent company Hon Hai Precision Industry Co. Ltd.) (1974). Precision engineering, nanotechnology, and computer components. Contract manufacturer for Apple, Microsoft, Google, Intel, Huawei, Sony, and Nokia. Market capitalisation: $US86 billion. Employs a staggering 803,000 workers.

In automobile manufacturing, computers are now central, running astonishingly sophisticated electronics. The three most important firms globally are:

- Toyota, Japan (1937). Toyota, Lexus, and Subaru vehicles, also banking. Market capitalisation: $US214 billion. Employs 370,000.
- Tesla, US (2003). Electric cars, solar panels and roof tiles, and batteries. Market capitalisation: $US464 billion (at which level it is the seventh-largest US company). Share value: $US500.
- Volkswagen Group, Germany (1937). Also includes Audi, Bentley, Škoda, Porsche, Lamborghini. Market capitalisation: $US84 billion. Employs 322,000.

Government security and intelligence agencies have made the concept of privacy virtually meaningless. Their surveillance capacity is growing exponentially, and they determine what citizens are entitled to know.

Artificial Intelligence is a subject of much specialist debate, but its implications have been barely discussed, let alone understood, in the public sphere. It is likely to displace some process workers in manufacturing—part of a long decline that began decades ago; will play an increasingly significant, and positive, role in medicine; and has some potentially alarming implications for a surveillance state.

The digital revolution, contrary to what might have been expected, has played a central role in dumbing-down political discourse and narrowing many people's understanding of the world and its problems. Social media and the more traditional forms of media—newspapers, radio, and television—have contributed to 'confirmation bias'. This has led to community disengagement, and at a time when Australia and the United States have the largest cohort of professionals in their history, engagement has turned inward, perspectives have narrowed and shortened, and open, informed debate has been stricken. This threatens the central principles of democracy.

The case of Julian Assange raises very serious issues, and illustrates this problem. He founded WikiLeaks in 2006 as a whistleblowing project in an attempt to counteract the huge and growing imbalance between the information collected about citizens without their knowledge and arbitrary restrictions on what citizens were allowed to know. The surveillance state, the world of Orwell's *Nineteen Eighty-four*, was just around the corner.

Democratic practice rests on the premise that citizens know what they are voting about and who they are voting for. If such information is denied, or massaged to conceal the reality, can we still call our nation a democracy?

In 2010, Assange published a video showing the killing of Afghan civilians by US troops in army helicopters in 2007, and later released classified emails secured from within the military by Chelsea (formerly Bradley) Manning and Edward Snowden. (Ironically, the 2019 raids by the Australian Federal Police on the

ABC and News Corp journalist Annika Smethurst were over reports of the killing of Afghan civilians by Australian troops.)

Manning was jailed, Obama later commuted the sentence, and Snowden escaped to Moscow. In 2016, Assange published thousands of unsecured emails by Hillary Clinton when she was secretary of state, and the timing was thought to have contributed to her defeat by Donald Trump. Russian hackers were alleged to have provided the emails.

Assange holed up in the Ecuadorian embassy in London, and was then held in an English jail. If he were to be extradited to the United States, he would face eighteen charges and a potential prison term of 175 years, which he would be unlikely to survive.

Alan Rusbridger, as editor of *The Guardian*, published material from WikiLeaks and wrestled with the moral dilemma involved, which he described in his important book *Breaking News: the remaking of journalism and why it matters now* (2018).

From one perspective, stealing secret government documents constitutes a major threat to national security, may endanger lives, and should be subject to severe criminal sanctions.

The alternative view is that governments routinely invoke 'national security' and classify routine documents as 'top secret' to avoid public disclosure or debate about matters of legitimate public concern, and are horrified by the prospect of being caught out for lying and wilful exaggeration.

The debate is unresolved.

Rusbridger writes that digital technology and the World Wide Web 'have created the most prodigious capability for spreading lies the world has ever seen'.

The digital revolution touches almost every aspect of our lives every day; the benefits of easy communication and access to knowledge are unprecedented; and clearly there is no turning back. But the digital revolution has been an integral part of the breakdown of accountability in public life, the destruction of the concept of the common good or public interest, the development of the surveillance state, psychological dependence on the technology itself, the speeding-up of everyday life, greater emphasis on consumption, the destruction of privacy, and the evolution of the citizen into raw material for exploitation.

The digital revolution has thus also become an instrument for nativist populism, conspiracy theories, the cherrypicking of evidence, and the trashing of expertise. This can be seen in the prominent example of the anonymous alt-right conspiracy-theory source known as 'QAnon' or 'Q', disseminated on the Internet. Adrienne LaFrance provides a long, forensic analysis of the organisation in the June 2020 issue of *The Atlantic*.[7]

QAnon, which emerged in 2017, is in the longstanding American paranoid millenarian tradition. Its central theme is to attack the Enlightenment and its legacies, asserting that the United States is run by a 'deep state', controlled by a cabal of Satan-worshipping paedophiles who dominate Hollywood. Donald Trump is a harbinger of 'The Great Awakening', who will use military power to destroy his enemies and bring in a millennium of peace, including the Second Coming. Use of the recurrent trope 'The calm before the storm' is a sign of recognition. It gives a sense of optimism to a segment of people who are baffled and frustrated by the modern world, and has made a significant penetration into the Republican Party.

Donald Trump has retweeted QAnon material about 150 times. If he wears a yellow tie or uses words starting with 'q', this is taken as an endorsement of QAnon.

QAnon asserts that Hillary Clinton was running a paedophile ring; the coronavirus pandemic and concerns about climate change are hoaxes; Barack Obama is planning a coup; Islam threatens the West; and all experts are part of the so-called deep state.

An Australian follower of QAnon massacred Muslims in two Christchurch mosques in March 2019. And he filmed his massacre for circulation on social media.

The digital revolution has transformed our lives, mostly for the good, and we cannot retreat from it. But the outreach of misinformation has increased exponentially. And Donald Trump can be regarded as its end product—so far.

CHAPTER FIVE

The Trump Phenomenon

In 2016, Donald John Trump succeeded, against very long odds, in what would have been described in business as a hostile takeover of the Republican Party. Among former Republican presidents or presidential candidates, George H.W. Bush, George W. Bush, and Mitt Romney did not vote for Trump in the November election, and a terminally ill John McCain left instructions that Trump was not to attend his funeral.

In July 2016, the controversial film-maker Michael Moore made the unpalatable prediction, against all the received wisdom of the commentariat, that Donald Trump would win the presidency. He offered several reasons, including 'Midwest Math, or Welcome to Our Rust Belt Brexit', pointing to blue-collar workers in traditional manufacturing states who saw themselves as victims of globalisation and warmed to Trump's aggressive nationalism and appeals to nostalgia; and 'The Last Stand of the Angry White Man', which could be summarised as, 'For 240 years white males have dominated the US. Then a black man was president for eight years. Now they want a woman! It's time to take a stand.'

In the campaign, when Trump exclaimed, 'I love the poorly educated!', it seemed both ludicrous and patronising, but he scored a palpable hit.

He taunted Hillary Clinton's supporters among professionals as 'elitists'. She then compounded her problem by ridiculing elements in Trump's base as 'deplorables'.

The Republican Party, curiously identified with the revolutionary colour red, holds sway in the American heartland, between the Rockies and the Appalachians (Illinois and Minnesota excepted), and in the formerly Democratic South.

Trump made very effective use of his campaign slogan, 'Make America Great Again' (#MAGA), in the areas where it mattered—middle America—away from the trade-exposed, more cosmopolitan great cities of the Pacific and Atlantic coasts.

'Again' appealed to nostalgia, back to an unspecified era sometime between the Middle Ages and black-and-white television.

In November 2016, he won 30 states and 304 votes in the Electoral College with 46.1 per cent of all votes cast. Hillary Clinton won 20 states and DC, with 48.2 per cent of the popular vote, a plurality of 2.86 million—but only 227 Electoral College votes.

Clinton won only three of the ten most populous states (California, New York, and Illinois) to Trump's seven (Texas, Florida, Pennsylvania, Ohio, Georgia, North Carolina, and Michigan).

Trump won in 2,626 counties, most of them rural and dispersed, while Clinton won in 487, almost all cities.

But there is a puzzling statistic that has to be recognised: there were 206 counties, with a total aggregate vote of 7.5 million, that voted for Obama in 2008 and 2012, but for Trump in 2016. So there has to be caution about overstating the significance of race as a factor in Trump's victory.

The states with the largest number of these 'pivot' counties were Wisconsin, Michigan, Pennsylvania, Ohio, Iowa, and Florida (the six states that, by changing the Electoral College vote, put Trump in the White House), and Minnesota, New York, Maine, Delaware,

Rhode Island, New Hampshire, and Maine, which remained with the Democrats.

There are other plausible explanations.

In 2008, Obama was an outsider who became a rather cautious insider in office, while long-serving Senator John McCain was more of an insider. (Mitt Romney, Obama's 2012 challenger, was an insider). In 2016, Hillary Clinton was the insider; Trump, an outsider. In 2020, Joe Biden was a classic insider, having served from 1973 to 2017 (44 years!) in Washington, while Trump, even as president, performed the role of outsider—at one point even urging voters in some states to rebel against their governors.

Surveys suggest that after Bernie Sanders, an outsider, lost the 2016 Democratic nomination to Clinton, about 14 per cent of his supporters voted for Trump.

In *Nervous States: how feeling took over the world* (2018), the English sociologist William Davies argues persuasively that appeals to 'nostalgia, resentment, anger and fear' have transformed democratic practice. He is writing essentially about the US and UK. In Australia, where compulsory voting ensures a high turnout, his thesis may be less compelling, but it deserves to be examined.

He offers an interesting explanation for why affluent old people and poorer white workers are now voting in the same way—for example, on Trump and Brexit—and he suggests two factors.

The first is hostility to immigration, 'neatly combining the desire for cultural continuity with anxieties about labour-market competition, and dividing the working class on ethnic lines'.

The second is physical pain: with the aged, the inevitable consequence of living into an eighth, ninth, or tenth decade, and with workers, a combination of factors resulting from poverty, poor nutrition, and inadequate medical and dental care. Davies estimates that about one-third of the community experience regular pain. It makes them unhappy, expressing deep anger, frustration, and disappointment. (They are also likely to be heavy users of analgesics.)

In June 2018, the *Journal of the American Medical Association* published a detailed study 'Association of Chronic Opioid Use With Presidential Voting Patterns in US Counties in 2016', indicating that 88 per cent of the counties with the highest rate of opioid dependence voted far more strongly for Trump in 2016 than they had for Romney in 2012, suggesting that pain was a major factor in changing votes. There is an extensive literature on this—also pointing to correlations between higher-than-average suicide rates and alcohol use.

The Electoral College system is weighted towards the Republicans, because two senators from states with very small populations could outweigh a majority of the popular vote of up to eight million resulting from crushing Democrat victories in states such as California, New York, Illinois, New Jersey, Virginia, Maryland, Colorado, Oregon, Minnesota, Washington, and Massachusetts.

Simon Schama distinguished between 'Godly America' with, at its heart, a farm, a church, and a barracks, and 'Worldly America … a city, a street, a port'.

During the 2016 campaign, Trump admitted that he had never read a biography of a United States president; indeed, that he had rarely finished any books, even the ones with his name on the title page.

The Trump phenomenon was, in part, a reaction to, and exploitation of, the digital revolution, with the use of Twitter as a primary form of communication. It also reveals a lack of trust in existing institutions.

He lives in a Trumpocentric universe, in which everything revolves around him. His self-absorption is the main driver of action, the concept of truth is irrelevant, evidence is discounted or dismissed as 'fake news', and gut reactions and instinct override analysis, evaluation, and the need to consult widely and take account of contrary views. Trump appealed powerfully to the self-identity of people who felt that that they had become marginalised and condescended to, whose views, or race, gender, patriotism, and consumption patterns could not be easily reconciled with modernity. It was central to his appeal. Gender, race, and sexuality raised disturbing questions, not publicly discussed but often internalised. Globalisation threatened employment in many areas, creating a sense of being left behind.

Greed, lack of empathy, rejecting altruism, dismissing expertise, attacking the weak, letting prejudice and racism flourish, dismissing the experience of other countries as irrelevant, generating fear and anger, were not regarded as political assets. They are now.

Trump's great strength was to reach out to the many millions for whom immediate self-interest, and small circles around it—family, friends, the locality, common interests (language, race, sport, religion, consumption patterns) and the short term—were more powerful motivating forces than appealing to 'the better angels of our nature'. Education was not an asset. Indeed, those pursuing the high moral ground, the long term, the universal, relying on expert knowledge, were detested as condescending. Presenting 'alternative facts' and dismissing material that could be subject to verification as fake news was central to Trump's appeal when millions responded: 'That's just the way I feel.'

Trump spoke directly into an echo chamber inside the heads of his followers, beyond reality, facts, or rationality.

Donald Trump habitually clapped himself in public appearances, saluted military personnel, despite having been conspicuously civilian all his life, praised himself without restraint (or embarrassment), and referred to himself in the third person ('President Trump has done more for the black community than any president since Abraham Lincoln.') These could be dismissed as mere eccentricities—but they may be pathological symptoms. He has been accused of being an habitual liar.

Politicians are often regarded as economical/parsimonious with the truth. But in Trump's case he may be incapable of distinguishing between truth and falsehood. He exhibits some of the characteristics of 'glossolalia'—'speaking in tongues'—with no gap between what is going on in his head and what comes out of his mouth. Sometimes he may surprise himself by what he has just said. He lacks any capacity to edit or impose quality control.

Mussolini and Berlusconi shared some characteristics with Trump: both were essentially showmen.

Trump's repeated refrain in the 2016 election, 'I will be your voice', had worrying echoes of Europe in the 1930s for people with a historical memory, but encouraged many without one. But the parallel cannot be pushed too far.

Hitler was quite different from Trump. Hitler had a coherent program—not sufficiently recognised until too late, and the capacity to carry it out (as the Holocaust demonstrated). Trump was not much interested in governing. His goal was to be the centre of the world's attention—and he certainly succeeded there. His politics is entirely personal.

Trump's program was himself, and the incoherence, lies, and distortions were built in. He would, characteristically, disseminate a falsehood by saying or tweeting, 'Some people say that Congressman X is guilty of incest, but I wouldn't know about that', a classic piece of 'dog whistling' where he dropped in a lie and immediately distanced himself from the consequences.

Trump is a sinister clown who inflicted serious damage to democracy, to judicial process, to rational debate, to global co-operation, and to the international standing of the US. Putin and Xi can congratulate themselves that Trump weakened the United States and advantaged Russia and China.

The most worrying factor about President Trump was his complete lack of curiosity. On the issues raised with him since his election, he either knew the answers already, or had no desire to hear the case for and against a proposition. He described himself as 'a very stable genius'.

He had the attention span of a humming bird and the emotional reactions of a greedy child. He had no concept of the separation of powers, the rule of law, scientific method, critical analysis, or the harnessing of expertise.

Typically, he surrounded himself with zealots such as John Bolton, General John Kelly, and Anthony Scaramucci, who then became bitter enemies and who reported that Trump hated briefing sessions and that the rare cabinet meetings (quite unlike the British or Australian models) were essentially monologues.

His political agenda, day by day, judging from his tweets, was determined, not by briefing from government agencies, but by the agenda on Fox morning television, especially the program *Fox & Friends*.

Trump was an early enthusiastic propagator of the accusation that Barack Obama had been born in Kenya, not in the United States, and was therefore ineligible, under the Constitution, to serve as president. There was a second accusation: that Obama was a Muslim. After his election as president, Trump grudgingly withdrew the 'birther' accusation, but it was still widely believed throughout his political base.

Bizarrely, the only international recognition that Trump appeared to covet was the Nobel Peace Prize, a side effect of his pathological envy of Obama, who received it in 2009.

Repeated polling confirmed that Donald Trump's most loyal supporters were religious fundamentalists—a mysterious finding, given his personal history, business ethics, exploitation of women, habitual lying, and shifty positions on matters of faith. Despite their misgivings about his personal morality, his declared opposition to abortion and the appointment of Supreme Court justices committed to overturn *Roe v. Wade* were vital to his support base.

The respected Pew Research Centre estimates (2020) that white evangelical Protestants account for 16 per cent of the US population. Eighty-three per cent of them lean towards the Republican Party, 89 per cent believe that the Bible should influence the laws, and 13 per cent think that 'God chose Trump'. This group was the core of President Trump's political strength. Rick Perry, a former governor of Texas, and later US secretary for energy, opined that Trump was 'the Chosen One'—another example of God using an imperfect human (together with Kings David and Solomon) to carry out his will.

A 2012 survey by the University of Chicago found that when a representative sample of Americans were asked, 'Does the Earth go around the Sun, or does the Sun go around the Earth?' only 74 per cent came up with the right answer. The survey found that 18 per cent of Americans reject evolution altogether, 48 per cent think that evolution was a God-driven process, and 33 per cent accept Darwin's theory of natural selection.

And yet United States citizens have won or shared more Nobel Prizes (383) than any other nation, by far.

Trump's rejection of scientific evidence about climate change and his withdrawal of the US from the Paris Accords on reducing global greenhouse-gas emissions was based on his obsessive conviction that 'the concept of global warming was created by and for the Chinese in order to make US manufacturing non-competitive'.

He loves to be loved, and exhibited a troubling enthusiasm for authoritarian rulers, many with blood on their hands, and was eager to spend time with them: Vladimir Putin, Xi Jinping, Kim Jong-un, and the Crown Prince of Saudi Arabia, Muhammad bin Salman. He showed disdain for democratic leaders such as Theresa May, Emmanuel Macron, Angela Merkel, and Justin Trudeau, and condescended to Malcolm Turnbull and Scott Morrison, even though he described the latter as 'titanium man'.

He accepted assurances by Putin and the Saudi crown prince that they were not involved in murders in England and Turkey, rejecting the intelligence-based conclusions of the CIA.

Describing Trump's operating method sounds amusing, but his narcissism, paranoia, and contempt for evidence has torn the social fabric, and threatened civil society.

Trump was insistent that his inauguration in January 2017 had drawn more people to Washington, DC, than for any other president, a demonstrably false claim.

His International Holocaust Remembrance statement in January 2017 failed to mention the Jews or anti-Semitism.

He told a rally in Florida in February 2017 that there had been a terrorist attack in Sweden. This was news to the Swedes. The explanation was that he had misunderstood a commentator on Fox News. He appointed a former coal lobbyist, Andrew Wheeler, to run the Environmental Protection Agency, and directed that government agencies were not to use the term 'climate change', and that the National Aeronautics and Space Administration and the National Oceanic and Atmospheric Administration were no longer to publish updates on climate-change issues.

Puerto Rico was devastated by Hurricanes Irma and Maria in September 2017, causing $95 billion in damages and 3,057 deaths. Trump responded with a grossly inadequate government provision of disaster relief, which he described as an 'A-plus' response. He said the actual death toll was sixteen, and visited Puerto Rico to hand out paper towels. Substantial funding did not arrive until 2019.

He argued that, before he took office, the US had the world's highest income taxes. It currently ranks no. 16, but, even before his sweeping tax cuts were introduced, many countries had higher tax rates, including Belgium, Denmark, Germany, France, Spain, the United Kingdom, Finland, Italy, Spain, Sweden, Israel, Canada, New Zealand, Japan, and even Australia.

After mass shootings in Charleston, Charlottesville, El Paso, Dayton, and Las Vegas, President Trump had seen no reason to take action on racial violence or gun ownership. He asserted that many racists 'are very fine people'. The First Amendment of the US Constitution, enshrining the right to bear arms, is the only part of the document that he has ever read. He insisted that the killings were essentially a mental-health problem, nothing to do with guns.

In March 2019, when an Australian white supremacist shot Muslims in two mosques in Christchurch, New Zealand, killing 51 and wounding 49, Trump's reaction was entirely predictable. True, he telephoned Prime Minister Jacinda Ardern—he would have been anxious to know whether she liked him—but his comments in a following press conference were appalling.

He admitted that he did not know much about what had happened in Christchurch, but he thought it was just another mental-health problem—an act by a deranged individual: racism did not come into it, nor the availability of guns (in the accused's case, an AR-15). So Trump, yet again, was giving another nod to his constituency, exactly as he did after similar events in the US: no need to change anything.

Many books, and countless articles and blogs, have been published about the decline of liberal values in many Western countries, including attacks on a free press, an independent judiciary, and the rule of law, and increasing resorts to secrecy and the rejection of pluralism. Vladimir Putin, in an interview with the *Financial Times* at the G-20 Summit in Osaka (June 2019), asserted that Western democratic values such as multiculturalism and social tolerance were no longer accepted by most people. An American journalist asked President Trump, at his Osaka press conference, to comment on Putin's claim that 'Western-style liberalism was obsolete'.

Trump's response was astonishing. He appeared to be unaware of the serious, continuing debate about the threat to Western liberal values, of which he was part. He thought Putin was talking about California (that is, the west from a purely United States point of view), and then engaged in a diatribe during which he agreed with Putin's non-existent criticism of city administrations by liberals in San Francisco and Los Angeles.

In April 2019 in London, speaking to British prime minister Theresa May before a business roundtable discussion, Trump said, 'We are your largest [trading] partner. You're our largest partner. A lot of people don't know that. I was surprised. I made that statement yesterday, and a lot of people said, "Gee, I didn't know that." But that's the way it is.'

Wrong and wrong. Britain's largest trading partner, at least for the time being, is Germany. The United States' largest trading partner is China. Why would he attempt to peddle a falsehood when May would obviously know what the truth was?

Trump told the German chancellor, Angela Merkel, that his father had been born in Germany. That was true of his grandfather, but Fred Trump, his father, was born in The Bronx. Why the lie? Of course, Trump would have not seen it as a lie, but as a piece of harmless flattery, telling a client—Merkel—what he thought she would like to hear.

On a later occasion, he told Merkel that she was 'stupid'.

He described a meeting with Queen Elizabeth II as 'incredible … we had automatic chemistry'.

He told Emmanuel Macron that he thought Bastille Day celebrated a French military victory. When he visited Hawaii in December 2017 for a commemoration of the Japanese attack on Pearl Harbor (1941), he appeared to have a very shaky grasp of its significance. He has not mastered the words of the 'Star-Spangled Banner'.

On 4 July 2019, President Trump said in his address to commemorate the Declaration of Independence, that in 1776, 'The Continental Army suffered a bitter winter at Valley Forge, found glory across the waters of the Delaware, and seized victory from [General] Cornwallis at Yorktown. Our army manned the air, it took over the airports, it did everything it had to do.'

Trump did not appear to grasp that the Civil War was not a celebration of the unity of the United States, but a bloody and potentially lethal attack on it: he seemed to see it as a celebration of the diversity of the greatest nation in the galaxy.

The Guardian's columnist Jonathan Freedland noted (on 12 July 2019) that if we were to see this as comical, it might distract us from the more important issue of Trump having turned the 4 July event—normally apolitical—into a military pageant, like those in Moscow, Pyongyang, and Beijing (although, to be fair, one could add 14 July in Paris). And a precedent had been set for 2020, a presidential election year.

In September 2019, the president tweeted that Alabama had a 95 per cent risk of being hit by Hurricane Dorian, a claim immediately denied by the National Weather Service. He then appeared on television with a weather map, showing an additional loop that had been made with a Sharpie pen, to confirm his assertion, and the National Oceanic and Atmospheric Administration was directed to amend its forecasts to indicate that Alabama could have been at risk.

His 53-minute telephone rant on *Fox & Friends* in November 2019, at a time when the House of Representatives was proposing his impeachment, contained eighteen false claims, in addition to his inflammatory attacks on individuals and his fixation with 'spying'. He said of his notorious call to Ukraine's President Volodymyr Zelensky, 'It was appropriate. It was perfect. It was nice. It was everything.'

In 2019, he launched into bizarre tirades about 'wind farms', which he described as 'windmills'; then, after saying 'I never understood wind', immediately contradicted himself and asserted that he was an expert on the subject. He blamed China and Germany for creating 'tremendous, tremendous amounts of fumes and everything ... spewing into the atmosphere'. Later, he engaged in an incoherent ramble about people flushing the toilet '10, 15 times', when once would suffice. He insisted that wind turbines were 'ugly' (in striking contrast to coal-fired power stations) and that their noise caused cancer.

In his speech to the United Nations General Assembly in September 2019, the president repeated the 'nativist' line and rejected international co-operation:

> Looking around and all over this large, magnificent planet, the truth is plain to see. If you want freedom, take pride in your country. If you want democracy, hold on to your sovereignty. And if you want peace, love your nation. Wise leaders always put the good of their own people and their own country first.
>
> The future does not belong to globalists. The future belongs to patriots ...
>
> Patriots see a nation and its destiny in ways no one else can.
>
> Liberty is only preserved, sovereignty is only secured, democracy is only sustained, greatness is only realized, by the will and devotion of patriots.

He made no reference to climate change, or even to international trade, and the coronavirus pandemic was yet to come. He did not make it clear how patriotism could solve global problems.

The president campaigned against the metric system for months in 2019, but a tweet on Christmas Eve confirmed that he did not understand what he was arguing against: 'Democrats are the metric-system party. They want to take your ounces and turn them into mega centimeters. Your inches aren't safe. They are coming for your feet. You won't be able to walk in kilograms.'

At least he did not call for abandoning dollars and cents; it may not have occurred to him that the US currency is based on a metric model.

Many voters felt very comfortable with his repetitive style, repeating claims and statements over and over again, manufacturing evidence, and plucking statistics off the wall—presumably because they interpreted and explained the world in the same way themselves.

The *Washington Post*'s Fact Checker calculated that, by August 2020, President Trump had made around 20,000 false or misleading statements since taking office, but his habitual lying and absurd exaggerations did not weaken support in his political base. It may even have strengthened it.

President Trump characteristically condemned the 'Black Lives Matter' protests after the death of George Floyd in Minneapolis as a challenge to law and order, and emphasised that 'when the looting starts the shooting starts'.

He caused outrage by using the National Guard and US Park Police to clear away crowds from Lafayette Park, in front of the White House, with some aggression, so that the president could have a 'photo-op' outside St John's Church, holding a Bible. The sheer hypocrisy of this startled even some of his conservative base, and was condemned by the Episcopalian and Catholic bishops of Washington.

Worse was to come. He gloated in a tweet about how 'the S.S.' had handled protestors 'very easily'. Did he really mean his own country's Security Service (which never uses those initials), or was it a Freudian slip, harking back to Hitler's Schutzstaffel or SS? (The president's father, Fred Trump, had been a member of the pro-Nazi America First organisation in the 1930s.)

The impeachment

The impeachment of President Trump and his subsequent acquittal by the US Senate in February 2020 set very dangerous precedents for democratic practice, the rule of law, the treatment of evidence, and even the nature of truth. Mitch McConnell's role as Republican majority leader in the US Senate had been deplorable in defying the constitutional principle of the separation of powers: refusing to schedule debate on 400 bills passed by the House of Representatives; and taking the leading role in stacking the judiciary and applying double standards, so that a nomination by President Obama to fill a Supreme Court vacancy in 2016 was denied 'because it is an election year', while Trump was assured that any nomination he made in 2020 would be acted on until October. McConnell's management of Trump's trial in the Senate after impeachment was simply a numbers game, a travesty, without witnesses, evidence, documentation, or cross-examination.

Each senator swore an oath 'that in all things appertaining to the trial of the impeachment of Donald John Trump, President of the United States, now pending, I will do impartial justice according to the Constitution and laws, so help me God'. However, McConnell made it clear that he was 'coordinating' with the White House counsel in the proceedings and that he would use his numbers to acquit. A two-thirds majority would have been required to convict Trump and remove him from office.

Senators then voted, on party lines, not to call witnesses and examine relevant documents. In the impeachment proceedings in 1999 against President Bill Clinton, a Democrat, excerpts from three videotaped depositions from witnesses were shown to the Senate, and there was extensive documentation.

Alan Dershowitz, a celebrated Harvard law professor (and registered Democrat), argued on Trump's behalf what is now called 'the Dershowitz doctrine': 'If a president did something that he believes will help get him elected, in the public interest, that cannot be the kind of quid pro quo that results in impeachment.' This gives a free pass to every politician who believes his re-election would serve the public interest and is prepared to lie or cheat to win. *The New Yorker* described this as '*L'état, c'est Trump*.'

With the Senate having acted as an agency of the executive, the acquittal destroyed the constitutional principle of the 'separation of powers'. The result normalised Trumpism, the authoritarian methodology, the rejection of expertise, the denunciation of enemies, contempt for alternative points of view, and impulsive, erratic, and ill-informed judgements.

More widely, the result signified that evidence, objective truth, and accountability no longer matter in politics—and if political leaders around the world habitually lie, exaggerate, or evade, that is part of the new rules of the game.

Chief Justice John Roberts, who presided over the impeachment process, described the United States Senate as 'the world's greatest deliberative body'. I assume he was being ironic.

Fintan O'Toole was on the mark when he described the process as 'an exercise in self-abasement', a refinement of the trial scene in *Alice in Wonderland*: 'verdict first, trial afterwards, sentence never'.[1]

O'Toole was puzzled that the Republicans did not even go through the pretence of having a show trial—even if the outcome was a fix, it would have looked better. Clearly, cynicism has reached such a point that it was not thought worthwhile to even make an attempt.

Trump's sophisticated defenders—and there were some—insisted that we should not confuse style and substance. His *modus operandi* was to seek attention and applause, even love, and his outrageous tweets did just that. Attacks on friendly governments, including the hapless Theresa May? What fun! The wholesale obliteration of enemy states? Well, we'll see.

Witty attacks by television satirists such as Stephen Colbert, brilliant articles in *The New Yorker* and *The New York Review of Books*, deep investigative reporting in *The New York Times* and *The Washington Post*, astute commentary on CNN, and countless sites accessed through the World Wide Web are likely to be lost on unemployed auto workers in Michigan and miners in West Virginia who once were Democrats and are now grumpy Republican voters.

As Simon Schama observed:

> A genuine sign that quasi-fascist policies are replacing democratic contestation is a preference for rallies over debates where rapture can be orchestrated, the press threatened (often physically), and where the Dear Leader leads the chants of 'Lock Her up'—the Trumpian remix of '*Sieg Heil*'. The fuel of these staged exercises in mass hysteria and verbal violence is the energy fascism exploited to great effect in the 1930s—the glee, the pure adrenaline rush of punitive hatred. The more horrified liberals become at these kinds of antics, the greater the surge of that hatred, juiced as it is by personal verbal abuse of the kind Trump has imported from reality TV and shock-jockery.[2]

The great casualty of the attrition of an independent critical press is, of course, the status of truth; the factual-knowledge project that, in the end, was the alpha and omega of the Enlightenment, with the ability to make critical judgements based on evidence.

A free press and investigative journalism were denounced as 'enemies of the people' by Trump. This was a French Revolutionary touch, unconscious in his case, but menacing.

When a free press is diminished or circumscribed, power is seized by interest groups, lobbies, faith healers, and shamans. Under Trump, failure by the United States to address the challenge raised by the science of climate change was jeopardising the safety of the planet and the life of future generations, just as the delays, confusion, and misinformation about COVID-19, and the failure to acknowledge growing racial tensions and deepening inequality, weakened the nation and crippled its moral authority.

Newton's Third Law

Isaac Newton published his Laws of Motion in *The Principia* in 1687. The third law, 'Every action has an equal and opposite reaction', seems curiously applicable, not only to physics, but also to history and the social sciences.

Everything is in motion. The tide retreats, then comes surging back.

Breaking down misogyny, racism, and fundamentalism; ending censorship; freeing up religious debate; and promoting multiculturalism often provoke a hostile reaction. The blowback includes the revival of white supremacy; the justification of hate speech and trolling as essential elements of freedom; the demonising of minorities; sharp attacks on expertise; and the denunciation of challenging evidence or reasoned argument as 'fake news'.

When Barack Obama won the presidency in 2008, defeating Hillary Clinton in the Democratic primaries and John McCain in the presidential election, it may have seemed reasonable to think (as many did) that, *The United States is entering a new era of hope. Electing an African-American as president means that the Civil War is at an end, white supremacy has been dealt a fatal blow, and the United States can pursue openness, optimism, and inclusiveness …*

It seemed inevitable that after ending a white male monopoly on the presidency, it was time to destroy another barrier—not race, but gender, by electing Hillary Clinton. The prospect of Donald Trump winning the Republican nomination in 2016 seemed far-fetched, and his election a virtual impossibility.

We know better now.

It is doubly ironic that for the 2020 presidential election, of twenty prospective Democratic Party candidates, the nomination was won by former vice president Joe Biden, another white male, even older than Donald Trump—with a patchy record on misogyny and race, a shaky grasp of history and language, and a recipient of campaign donations from the fossil-fuel industry.

Inevitability and eternity

In *The Road to Unfreedom: Russia, Europe, America* (2018), Timothy Snyder proposed a compelling hypothesis that democracy is threatened by two types of deterministic worldview: 'inevitability' and 'eternity'.

After the fall of the Soviet bloc, the establishment of a kleptocracy in Russia, and China's aggressive pursuit of state-directed capitalism, neoliberal economics was the only viable system operating globally. As Margaret Thatcher used to insist, 'There is no alternative', and the acronym 'TINA' was often used. Essentially, TINA represented a top-down imposition by a technocratic elite. Open debate and serious questioning were no longer relevant. And if you lost your job in a rustbelt area, well, tough luck: nothing could be done about it (apart from welfare support). Hillary Clinton's endorsement of continuity and 'inevitability' was a major factor in her losing the votes of blue-collar workers and their families, historically supporters of the Democrats.

Trump's threat of a wrecking ball proved to be powerful with voters who felt they had become victims of the triumph of neoliberal globalisation: he could reverse the inevitable.

'Eternity' is all about mythic nationalism, taking a very long view; in the case of the United States, back to the first European settlement in the seventeenth century.

According to this view, Americans are essentially white, European, English-speaking, and Christian, in a society dominated by males, and First Nations and African slaves are peripheral to the national narrative. Even events such as the US Civil War are only minor perturbations in the context of 400 years of white occupation.

Vladimir Putin's view of Russia is dramatically different from Vladimir Lenin's—or Mikhail Gorbachev's—based on a historic culture shaped by a millennium of isolation, religious hierarchy, and Orthodoxy. The period of communist rule from 1917 to 1989 could be regarded as an aberration.

Adam Tooze, in reviewing Snyder's book and other works about democratic threats, wrote:

> Our current situation … has been shaped by the wild oscillation between the determinism of modernization theory and the determinism of nationalism. Both foreclose any real debate and all practical alternatives. They are both inimical to genuine democracy. One licenses domineering technocracy; the other, cruder forms of authoritarianism.[3]

Trump's continued invocation of 400 years of white, Anglophone American history was highlighted by his ambitious posturing under the images of Washington, Jefferson, Lincoln, and Theodore Roosevelt at Mt Rushmore in South Dakota on 4 July 2020. His insistence that Confederate monuments had to be celebrated as part of the American dream epitomised his ignorance of and disdain for global problems, his rejection of international collaboration, and (as with COVID-19) his dismissal of evidence, expertise, and scientific method.

But it is his refusal to address climate change that may be the worst single element in his legacy.

CHAPTER SIX

Climate Change: the science

Throughout much of the world, and particularly in Australia, the climate is changing at a scale and speed unprecedented in modern times. The term 'global warming', especially the combination of higher temperatures and drought, is relatively easy to demonstrate and explain. But there are many other complex and diverse phenomena—extreme weather events, including floods, cyclones, and even some cooling—that are driven by global warming.

'Climate change' is a convenient, if understated, label that has been adopted by the Intergovernmental Panel on Climate Change (IPCC), treating it as a secular event in which 'anthropogenic'—human-induced—activity is a major driver. The evidence of global warming, with an ever-increasing global mean surface temperature, is now stark: the 2010–19 decade was the hottest on record, according to the World Meteorological Organization; the past two decades included eighteen of the twenty warmest years since international record-keeping began in 1850; and January 2020 was the hottest month on record globally.

The effects include:

- the melting of Arctic sea ice, and the cracking and melting of ice sheets in Antarctica and Greenland;
- the risk that warming will release methane clathrates (a lattice that traps molecules) in the Siberian tundra, the poles, and on ocean floors;
- the retreat of glaciers in the Himalayas, the Andes, the Alps, the Rocky Mountains, and Mt Kilimanjaro;
- the warming, deoxygenation, and acidification of oceans;
- sea-level rises;
- the spread of ocean dead zones and algal overgrowth;
- the destruction of underwater kelp forests, essential for fish breeding and the absorption of carbon dioxide;
- the mass deaths of corals;
- the contraction of agricultural land in North Africa and much of the Middle East; and
- the increasing frequency and severity of many extreme weather events (including more frequent devastating tornadoes and cyclones).

Adverse local events are unlikely to be the result of natural conditions alone. For example, in southern Africa, one of the world's largest waterfalls, the Victoria Falls, 'Mosi-oa Tunya' ('The Smoke that Thunders'), has been reduced to a trickle; and in Venice there have been extreme fluctuations of water levels, with record floods in November 2019, but the canals at a record low in February 2020.

The deoxygenation of oceans and an unusually high rate of species extinction are indirect results of climate change. Pollution from phosphorus and nitrogen enhances plant (algae) growth, which consumes oxygen in waterways; and as rainfall contracts, farmers are forced to clear woodlands, destroying the habitat of rare species.

The Amazon basin used to be described as 'the lungs of the world', absorbing about 5 per cent of the world's carbon dioxide.

However, since the 1990s this capacity has fallen by about 30 per cent: the rapid growth of trees has led to their early maturity and higher vulnerability. Also, high rates of forest clearing and burning off for grazing—probably illegal, but covertly encouraged by the Bolsonaro regime—means that the Amazon basin's contribution to carbon dioxide absorption is being sharply reduced.

Alexander von Humboldt recognised the impact of deforestation on climate in Venezuela as early as 1800.

And in Australia, the direct and indirect effects of climate change include:

- surface temperatures having increased by 1.5°C since 1900;
- 2019 being the hottest year on record, with the lowest recorded rainfall for 120 years;
- bushfire seasons beginning earlier—in July, in some states, even during winter—and now last longer, are more intense, and spread across a greater area, including subtropical rain forests (186,000 square kilometres, equivalent to 80 per cent of the area of Victoria, were burnt out in the summer 2019–20 bushfires);
- the bleaching of the Great Barrier Reef;
- increases in some extreme weather events;
- extreme drought and flooding—linked to the El Niño–Southern Oscillation (ENSO);
- polluted river systems;
- a decline in ocean fish stocks;
- the disappearance of bees and butterflies in many areas;
- reductions in the numbers of birds; and
- hazardous air quality in cities.

Record fish deaths in Australian rivers have been compounded by political decisions about water allocations, which give low priority to environmental needs. Species extinction becomes collateral damage.

Humans are used to living with dramatic changes in temperature. Places such as Melbourne have a temperature range of almost 40°C between summer maxima and winter minima, and year-to-year variations of average temperature of a couple of degrees.

But nature is extremely sensitive to small *aggregate global* changes. The biota may be extremely vulnerable to what seem trivial temperature variations (2–3°C) to humans.

Australia seems to be emerging as an outstanding case-study of the effects of climate change if no action is taken to reduce greenhouse-gas emissions. Scientists warned us about this decades ago.

However, far from leading worldwide efforts on climate change action, Australia's response has been feeble, confused and, at times, corrupt, obstructing attempts to secure binding international commitments to reduce carbon dioxide emissions.

The central elements of climate change/global warming

The atmosphere has five layers, of which the closest to Earth are the troposphere and the stratosphere. The troposphere surrounds the Earth's surface, with an average thickness of 13 kilometres (lower at the poles, highest in the tropics). The stratosphere is the next layer out, and we pass through it in long distance high-altitude flying. Further out are the mesosphere, the thermosphere, and the exosphere, the last extending up to 10,000 kilometres from Earth's surface.

The Earth's climate is, and always has been, changing due to many factors, including fluctuations in the incoming ('short-wave') radiation from the Sun and changes in the greenhouse effect through which water vapour, carbon dioxide, methane, nitrous oxide, and some other gases, known collectively as greenhouse gases, interact with the outgoing ('long-wave') infrared radiation, keeping the Earth's mean surface temperature about 33°C warmer than it would otherwise be.

The emission of greenhouse gases due to human activity is a major contributor to recent climate change/global warming ('anthropogenic climate change').

Since the last great Ice Age, which ended about 11,700 years ago, the global mean surface temperature (GMST) has increased by about 5.0°C, and the ocean temperature by almost 3.5°C.

Within that 5 per cent increase, the GMST has increased by 1.0° C since the period 1850–1900. That is a global average: in 2019, Australia was about 1.5°C warmer. With the current growth rate, average surface temperatures could rise above 2.0°C between 2030 and 2050, and a further 2.0°C by 2100, according to the US National Oceanic and Atmospheric Administration.

Carbon dioxide and other greenhouse gases stay in the atmosphere for a century or longer, except for unburned methane and a few other minor gases, absorbing and retaining more heat from the Sun.

The 'greenhouse' phenomenon has been recognised and its basic physics broadly understood since the nineteenth century.

Burning coal is the biggest single contributor to increased levels of carbon dioxide, accounting for 46 per cent of the total. The abundance and cheapness of coal exacerbates the problem. Methane leaks and the carbon dioxide that comes off at 'extraction' are major problems for fracking and gas fields.

Each tonne of carbon in coal, when burnt, produces 3.67 tonnes of carbon dioxide; petrol, 2.2 tonnes; and natural gas (methane), 1.9 tonnes. Fabricating each tonne of cement releases 0.9 tonnes of carbon dioxide.

Carbon dioxide levels in the atmosphere (estimated at 417 parts per million in May 2020) are now at their highest point for at least 800,000 years. Peaks and valleys in the chart below track the coming and going of ice ages (low carbon dioxide) and warmer interglacials (high carbon dioxide).

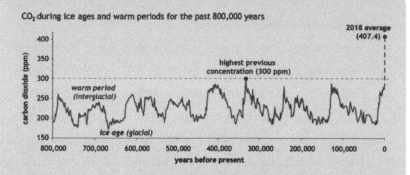

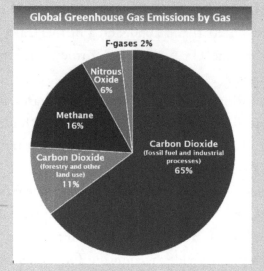

Livestock emissions (largely methane and nitrous oxide) account for about 18 per cent of greenhouse gases, added by human activity. With beef cattle, the equivalent of 300 kilograms of carbon dioxide is produced for each kilogram of protein.

Methane remains in the atmosphere for a far shorter period than carbon dioxide, but traps solar heat 84 times more effectively. 'Fugitive' methane accounts for 2 per cent of the emissions.

Source: Top image: NOAAS Climate.gov; Bottom image: Commons: Wikimedia.org

When the emission of greenhouse gases to the atmosphere exceeds the rate of absorption by the terrestrial biosphere and ocean, atmospheric concentrations must be expected to rise, leading to a dangerous interference with the global climate system.

The use of fossil fuels in transport, deforestation, massive land-clearing, waste-dumping in oceans, urbanisation, and construction, led by increased population and longevity, and compounded by per capita resource use, are the major contributors. Cement accounts for 8 per cent of the total—much more than the aviation industry (2.5 per cent).

An annual increase of more than 40 billion tonnes (gigatonnes: Gt) of greenhouse gases due to human activity has disturbed the near-equilibrium of pre-industrial times between their naturally occurring production and absorption, in which oceans and forests have acted as 'sinks'.

Increased carbon dioxide emissions and greenhouse warming leads to rising temperatures in the atmosphere and the oceans, polar ice-cap melting, and marine acidification and deoxygenation, but also other climate anomalies and more frequent severe weather events, including extremes of heat and cold, flood and drought, and hurricanes and tornadoes.

Climate change is a major factor in the destruction of habitat and species loss, especially bees, birds, and insects, with their major impact on agriculture.

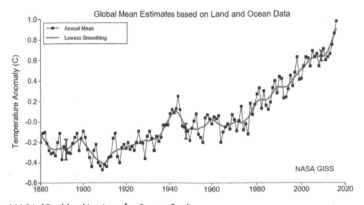

Source: NASA/Goddard Institute for Space Studies

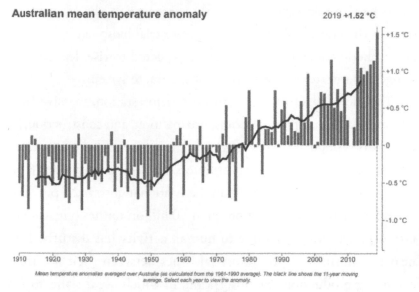

Australian mean temperature anomaly   2019 +1.52 °C

Mean temperature anomalies averaged over Australia (as calculated from the 1961-1990 average). The black line shows the 11-year moving average. Select each year to view the anomaly.

Source: Australian Bureau of Meteorology, 2019

In 2018, according to estimates by the International Energy Agency (IEA), carbon dioxide emissions amounted to 33.1 billion tonnes, the equivalent of 4.4 tonnes for each person on Earth, an increase of 1.7 per cent on 2017 figures. Emissions fell in Europe and Japan, but grew in India, China, the United States, and Australia.

Climate change as 'a diabolical problem'

In *The Garnaut Climate Change Review* (2008) commissioned by the Commonwealth government and the states, Professor Ross Garnaut, an eminent economist who had been Bob Hawke's economic advisor and became ambassador to China, described climate change as 'a diabolical problem'.

Coal burning produces carbon dioxide, and is the greatest single factor in driving unsustainable climate change.

We can't live without carbon dioxide because, as Antoine Lavoisier demonstrated in 1782, respiration is a form of combustion for humans, like the animals we live on or with. Forests, crops, and vegetation generally depend for growth on their absorption of carbon dioxide What we release in the atmosphere is the excess not taken up by plant life.

Much of the developing world depends on burning coal for light, heat, and cooking. In 2006, US coal lobbyists produced a television advertisement with a tag line about carbon dioxide: 'They call it pollution. We call it life.' It was, at best, a half-truth.

The central problem in reducing the production of carbon dioxide is that it intersects with most aspects of modern human life. This can be contrasted with stratospheric ozone, where a few billion dollars solved the problem (and China is now stopping the rogue production of ozone-depleting substances).

The medium-to-long-term risks are disturbing:

- Warming is causing the Siberian, Canadian, and Alaskan tundra to thaw, and if the poles and ocean floors are disturbed, this could release vast amounts of methane, far more potent as a greenhouse gas than carbon dioxide;
- If large areas of the Antarctic ice sheet and Greenland were to melt, it would lead to significant sea-level rise and risk drowning major urban coastal cities and towns; and
- Since 1954 there has been a growing recognition of the impact of deforestation on rising carbon dioxide levels, and since 1957 some evidence that seas were becoming less effective as carbon dioxide sinks than had been assumed.

Comparing Mercury, Venus, the Earth, and Mars is instructive. Mercury's average distance from the Sun is 57.9 million kilometres, and its surface temperature ranges from 430°C in daylight, down to −180°C on the dark side. Mercury has no atmosphere as we would define it.

Venus is much further out, on average 108.4 million kilometres from the Sun. But its surface temperature is hotter than Mercury's, 467°C in daylight, and an average of 462°C, day and night. The reason is that, unlike Mercury, Venus has an atmosphere, 96.5 per cent of which is carbon dioxide, with a surface pressure 9.3 times greater than Earth's. The carbon dioxide retains the radiant heat of the Sun and distributes it evenly.

The Earth, 149.6 million kilometres from the Sun, on average, has a global mean temperature of 15°C, where water vapour and other greenhouse gases comprise a tiny, but significant, percentage of the total mass of the atmosphere.

Mars' orbit averages 228 million kilometres from the Sun, and its mean temperature is −80° C. Its atmosphere is rich in carbon dioxide but with only 1 per cent of the Earth's pressure.

The amount of carbon dioxide present (or absent) in the atmosphere of each planet is decisive in determining whether cellular life can exist—let alone more exotic forms such as human life.

The drivers of climate change

Nine factors have determined climate change in the life of *Homo sapiens*. Three are beyond human influence (BHI), but our species makes a major contribution to three others, and has a possible connection to three more.

Variations in the Earth's orbit around the Sun

The most important factors in the very long-term variability of the Earth's climate are determined by changes in three cycles: the Earth's eccentricity in orbit, over a period of 100,000 years; the Earth's axial tilt relative to the Sun, 41,000 years; and precession, or changes in orientation, of the Earth's axis, 25,771 years.

'Milanković cycles' (1941) were named for Milutin Milanković

(1879–1958), a Serbian mathematician and geophysicist whose work only became generally accepted in the 1970s. These cycles have been decisive for the last two million years in determining climate, and are predictable but far too long for human society to adjust to. Humans, animals, and plants can adjust to change over long time scales where the transition is low and incremental. The problem with anthropogenic global warming is its rapidity. Geologists think in terms of millions of years, but most of us do not. Modern politicians are preoccupied with the next election. Psychologically, most people cannot take on a 50-year projection, let alone a millennium or two. (BHI)

Sunspots

There is an eleven-year 'solar magnetic activity' cycle marked by sunspots and solar flares that has a direct impact on the world's climate. The 'Little Ice Age', described below, had an exceptionally low number of sunspots ('the Maunder Minimum'). Although the Sun now appears to be brighter than it was a century ago, the level of sunspot activity has plateaued since the 1960s, falling slightly between 1986 and 2008. The impact on climate has been relatively limited, compared with anthropogenic changes. (BHI)

Volcanic eruptions

They are largely unpredictable. The greatest volcanic eruptions in recent centuries, on Mt Tambora (1815) and Mt Krakatoa (1883), both in Indonesia, had a global impact on climate by hurling huge volumes of particulate matter, notably sulphur, into the atmosphere, screening out the short-wave radiation from the Sun. The persistence of aerosol particles depends on the height to which they are ejected. In Europe, 1816 was described as 'the year without a summer', and the 1810s was the coldest decade on record. After Krakatoa, sunsets were darkened for a year in Australia and elsewhere. The eruption of Mt Pinatubo (1991), in the Philippines, which generated a 35-kilometre-high ash cloud, reduced temperatures in parts of the world for two years. However, when Mt Eyjafjallajökull, in Iceland, erupted in 2010, it did not send particulate matter up to the stratosphere, and no cooling followed. Climate-change denialists often assert that volcanoes produce more carbon dioxide than human activity, but the US Geological Survey calculates that in most years the ratio is 130 (human) to 1 (volcanic). (BHI)

Radiative forcings

The IPCC defines radiative forcing as the difference between insolation (that is, sunlight) absorbed by the Earth and energy radiated back to space. Changes to Earth's radiative equilibrium, causing temperatures to rise or fall over a period of decades, are called climate forcings. Radiant energy from the Sun (infrared, visible, and ultraviolet) delivers up to 1,360 watts per square metre: about half is absorbed by the Earth, the other half emitted back to the atmosphere as long-wave (infrared) radiation. Greenhouse-gas molecules also absorb and re-emit the infrared radiation.

The long-wave radiation, together with other physical processes—widely, but inaccurately seen as analogous to a garden greenhouse—have produced further surface warming of more than 30°C generally. This is the central problem in climate change, especially global warming, but rarely discussed and poorly understood. Infrared absorption matches the energies associated with bending the angles between bonds in a molecule, for which at least two bonds, and thus at least three atoms, are needed, so dioxygen (O_2) and dinitrogen (N_2) are not greenhouse gases. (Partly BHI)

Albedo

'Albedo' (Latin for 'whiteness') is the measure of the reflection of solar (short-wave) radiation. Fresh snow or cumulus clouds reflect about 70 per cent of solar radiation; ice, about 40 per cent; deserts, about 30 per cent; but forests, only 10 per cent. The oceans have a low albedo, around 8 per cent, and absorb very large amounts of solar radiation. The average albedo of planet Earth is 30–35 per cent, largely due to cloud cover.

However, humans can make a contribution to albedo. In the period 1940–70, the amount of particulate matter in the atmosphere in the Northern Hemisphere increased significantly, contributing to a slight cooling. 'Smog' (a combination of 'smoke' and 'fog') blocked out the Sun in many cities. World War II was a major factor, and clean-air regulation was either inadequate or absent throughout the period. The (sulphuric) acid-rain phenomenon and other toxic substances caused major health problems. Eight thousand deaths were attributed to the 'Great Smog of London' in 1952. Richard Nixon brought in clean-air legislation in the US, and the situation improved significantly in the 1970s.

There was a slight cooling in the Northern Hemisphere in 1940–70, but not in the Southern Hemisphere. Projecting reflective particulate matter, such as sulphur, into the stratosphere, a form of 'geo-engineering', might help to lower global temperatures, but the subject is controversial. (Partly BHI)

Regional weather cycles

Four very important cycles originate in the Pacific, Indian, and Antarctic Oceans: the El Niño-Southern Oscillation (ENSO); the Southern Annular Mode (SAM); the Indian Ocean Dipole (IOD); and the Interdecadal Pacific Oscillation (IPO).

The most significant is ENSO, which has two phases—El Niño ('the Christ child' in Spanish, named by Peruvian fishermen because the phenomenon usually occurred around Christmas), marked by an upwelling of warm water in the Central Pacific, and La Niña ('the girl'), by a cold upwelling. There is a persuasive argument that a very powerful and lengthy ENSO in 1788–96 led to devastating crop failures and starvation in France, and precipitated the revolution in 1789.

SAM, also known as the Antarctic Oscillation (AAO), describes the north-south movement of the westerly wind belt that circles Antarctica. If the westerlies move northward, the chance of cold fronts, rain, and storms in southern Australia increases; if they contract southward, rainfall diminishes.

The IOD, which was only identified by climate researchers in 1999, refers to sea-surface temperatures in the Indian Ocean and their impact on monsoons and rain in Australia. A positive IOD sees warmer water in the western Indian Ocean, with greater precipitation on surrounding continents. When the IOD is in its negative phase, with warmer water off north-western Australia, moisture is picked up from that ocean and delivered as higher rainfall in northern, central, and south-eastern Australia. The positive phase brings with it droughts in eastern Australia. This has a much more significant effect on rainfall patterns in south-eastern Australia than ENSO. Coral cores from the eastern and western Indian Oceans have allowed the construction of a Dipole Mode Index going back to 1846. This suggests that the positive IOD patterns—that is, those with links to Australian droughts—increased in strength and frequency during the twentieth century.

The IPO affects the whole Pacific, and the cycle varies between 15 and 30 years: the warmer seas become cooler; the cooler seas, warmer. Some research suggests that rising temperatures due to changing atmospheric composition resulting from greenhouse-gas emissions may be a factor in more intense El Niño events, but the issue is contested.

El Niño is a major influence on drought in our region, with 2–7-year cycles in the Pacific Ocean. They are still difficult to predict. It is now thought possible that human influence may have some impact, with a risk that ENSO-type conditions could become permanent. There have been about 30 El Niño events since 1900, the strongest being in 1982–83, 1997–98, and 2014–16. The very hot year of 2019 had a weak El Niño. (BHI?)

'Tipping points': loss of equilibrium

For most of the past 800,000 years there was a fine balance in the atmosphere between the production of carbon dioxide and methane—the sources (burning wood and, later, fossil fuels, respiration and belching by animals and humans), and the sinks (atmosphere, trees, plants, oceans). However, at no time did carbon dioxide exceed 300 parts per million (ppm) in the atmosphere. Industrial exploitation changed the impact of the carbon cycle, releasing in decades carbon (coal and oil) that had been laid down over millions of years, in the process seriously disturbing the environmental balance. Forest clearing, huge increases in population, and *per capita* consumption are tipping points, although emissions comprise a very small proportion of the constituents of the atmosphere. With carbon dioxide now in excess of 400 ppm, and rising, the risks are considerable that a 'tipping point' will occur and that warming may become irreversible.

Threats to equilibrium/climatic destabilisation, however described, create a variety of micro-climates, with floods *and* drought, heat waves *and* record snow falls, glaciers melting *and* some glaciers cooling. This is all harder to explain, probably by an order of magnitude.

Anthropogenic (human) emissions of greenhouse gases

Population growth, and greater longevity, higher per capita resource use, electricity generation, motor vehicles, aviation, changing land use, forest clearing, grazing, and waste dumping in oceans are all contributing to an exponential increase in greenhouse-gas production. The length of the carbon dioxide cycle is deeply controversial. Australian-born theoretical ecologist and former president of the Royal Society, Lord May (1936–2020) emphasised the century-scale response of the natural processes that disperse carbon dioxide in the atmosphere, while Anglo-American mathematical physicist Freeman Dyson (1923–2020) described a twelve-year timescale that could be modified, he thought, by the large-scale geo-engineering of vegetation. This 'independent' and added variable of human-related activity has skewed the normal balance by adding massive amounts of heat energy to global systems, the land atmosphere, and oceans.

Carbon dioxide is far more abundant in the atmosphere than methane or nitrous oxide, but both have a higher global-warming potential at the molecular level. Water vapour is the most important greenhouse gas, and increases mainly because warmer air can hold more water as vapour. But humans cannot modify the production of water vapour.

Ozone depletion

The major depletion of stratospheric ozone, especially above the polar regions, was attributed to the use of 'halocarbon compounds', manufactured chemicals that included chlorofluorocarbons, used as an aerosol propellant, refrigerants, and solvents, which catalysed the breakdown of ozone into oxygen. These were all greenhouse gases. The ozone layer provides protection against ultra-violet radiation, which contributes to cancers, sunburn, and cataracts. In 2003, the IPCC calculated the volume of halocarbon compounds released in the troposphere at around 200 ppt (parts per trillion), but concluded that they had an important role in 'radiative forcing'. The Montreal Protocol (1987), agreed on without dissent, phased out the use of several halocarbon compounds that, as well as being ozone-destroying, were powerful greenhouse gases. The ozone hole should be fully healed by 2060.

How our understanding of climate-change science evolved

The 'Medieval Warm Period' and the 'Little Ice Age'

In the very long term, climate (essentially the long-term aggregate of 'weather') is subject to a variety of causes, and in the last millennium these produced two enormous anomalies, now known as the 'Medieval Warm Period' and the 'Little Ice Age'.

The term 'Little Ice Age' was coined by François Matthes, an American geologist, in 1939, and the term 'Medieval Warm Period' was in use from the 1960s. Our knowledge of the earlier period is partly recorded in documents, but also stems from the study of ice cores, tree rings, and lake and ocean sediments. The later period was also recorded in art (in many paintings of icebound cities and rivers) and photographs.

The Medieval Warm Period, which ran from about c. 950 CE to c. 1250, was particularly significant in the North Atlantic region, probably due to increased solar activity and changes in ocean circulation. It appears not to have been global, and had little impact in the Southern Hemisphere.

Climate-change denialists or confusionists get quite excited about the Medieval Warm Period. Although bureaus of meteorology in the eleventh century were perhaps not as technologically sophisticated as ours, it is clear that temperatures were far higher than in the fifteenth, sixteenth, or seventeenth centuries in Europe and Asia. But that was *then*. We are dealing with *now* and *the future*, our lifetimes and the next generations'. The world's population in the year 1000 has been estimated at about 300 million. Our problem is dealing with the next century or longer, with thirty times the population of the medieval era, and with far higher *per capita* resource use. We are not comparing like with like.

Reconstructed Temperature

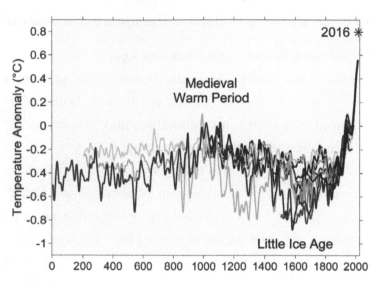

Source: Commons: Wikimedia.org

The Little Ice Age ran from around 1300 to around 1870, with its low point at around 1650. It has been attributed to several factors: decreased solar radiation; changes in ocean circulation; the degree of glaciation; massive volcanic eruptions after 1257; and falls in population due to the Black Death and colonial slaughter in the Americas. Its impact was probably greatest in Europe and North America. (In some countries, the Jews were blamed.)

Capturing radiant heat from the Sun

Horace-Bénédict de Saussure (1740–99), a Swiss from Geneva, experimented (1767) with glass 'hot boxes', which retained solar heat, anticipating research on the 'greenhouse effect'. He developed a solar oven.

The law of the conservation of mass

The discovery that total mass remains constant even after transformation by physical or chemical changes, such as burning (1774) is generally attributed to Antoine Lavoisier (1743–94).[1] However, the discovery had been made earlier—independently, in Russia, in 1748—by Mikhail Lomonosov (1711–65). In 1772, Lavoisier turned his attention to combustion, in one of his greatest contributions to science.

When burnt, coal is converted, as atoms of carbon and oxygen combine, into carbon dioxide, other gases, and particulates, but the total mass of carbon is unchanged. Each tonne of carbon, by burning, produces 3.67 tonnes of carbon dioxide. The total mass of carbon remains in the atmosphere (troposphere > stratosphere) and does not disperse significantly for about a century, perhaps longer.

It can be safely asserted that the science is settled on the conservation of mass, but not quite in the way that Lavoisier had in mind, given the challenges of $E = mc^2$, general relativity, and quantum mechanics.

How the atmosphere retains heat from the Sun, making life possible on Earth

Joseph Fourier (1768–1830), the great French mathematician, influenced by Saussure's pioneering work, hypothesised in 1824 that the atmosphere trapped heat from the Sun, more like an insulating blanket than a glasshouse, and that without it, the Earth's distance from the Sun would make it far too cold to support life. He anticipated the 'greenhouse effect', without using the term. Claude Pouillet (1790–1868) later calculated that without the atmosphere, the mean temperature on Earth (then 14°C) would be 32 degrees lower (–18°C). The roles of water vapour and carbon dioxide in absorbing solar radiation are central.

This discovery was first published in an important paper,

'Circumstances affecting the heat of the sun's rays' (1856) by Eunice Newton Foote (1819–88), an American scientist, inventor, and painter. She anticipated Tyndall's more detailed work, published in 1861: he was probably unaware of her discovery.

The identification of gases that absorb and emit solar radiation

John Tyndall (1820–93), an Anglo-Irish physicist, identified water vapour, carbon dioxide, methane, nitrous oxide, and ozone as gases that absorb and emit radiation within the thermal infrared range. He published his results in 'On the Absorption and Radiation of Heat by Gases and Vapours, and on the Physical Connexion of Radiation, Absorption, and Conduction' (1861), and gave public lectures on the subject. The gases he identified were later called 'greenhouse gases'; although they constitute only a tiny fraction of the total atmosphere, their role is decisive.

Calculation of 'the greenhouse effect'

Svante Arrhenius (1859–1927), a Swedish physical chemist, won the Nobel Prize for Chemistry in 1903. In 1896, working with Arvid Högbom, he calculated that doubling carbon dioxide levels in the atmosphere would raise surface temperatures by 4°C, and between 8°–9°C at the poles. A quadrupling would raise surface temperatures by 8°C. Because of his concerns about a future Ice Age, Arrhenius welcomed the prospect of a 'hot-house' effect, which he thought would result in a more benign climate in Northern Europe and higher crop yields 'in a few centuries'. In 1901, the 'hot-house' effect was renamed 'the greenhouse effect' by his colleague Nils Gustaf Ekholm.

Interest in Arrhenius's work was revived by the English engineer Guy Stewart Callendar (1898–1964), an amateur meteorologist, who calculated that the 'tipping point' for irreversible increases in temperature might be a further 2°C. His work influenced Dave Keeling, and the 2°C figure was later adopted by the IPPC.

'Living on our capital'

The prodigious American statistician Alfred James Lotka (1880–1949), in *Elements of Physical Biology* (1925), described what we now call 'anthropogenic climate change', but without using the term, a century after Fourier's work:

> Economically we are living on our capital; biologically we are changing radically the complexion of our share in the carbon cycle by throwing into the atmosphere, from coal fires and metallurgical furnaces, ten times as much carbon dioxide as in the natural process of breathing. How large a single item this represents will be realized when attention is drawn to the fact that these human agencies alone would, in the course of about five hundred years, double the amount of carbon dioxide in the entire atmosphere, if no compensating influences entered into play.

Lotka referred to 'the present régime of "evaporating" our coal mines … into the air', adapting the words of Arrhenius. He was optimistic in predicting that the doubling would take 500 years. In 1925, carbon dioxide constituted 305 ppm of the atmosphere: in the 95 years since then, it has risen by 35 per cent.

In effect, in a very short time, we have been releasing massive amounts of stored energy laid down in the earth (as oil, gas, and coal) over millions of years. Apart from anything else, is burning the best use we can find for this valuable resource?

Thomas Edison said, 'Sunshine is a form of energy, and the winds and tides are manifestations of energy. Do we use them? Oh, no; we burn up wood and coal, as renters burn up the front fence for fuel. We live like squatters, not as if we owned the property.'

In *Superpower* (2019), Ross Garnaut draws attention to a perceptive contribution by the Anglo-Australian economist Colin Clark (1905–89) in his *Conditions of Economic Progress* (1940). Clark appeared to be channelling Lotka in warning that we must not set out to burn up fossil fuels 'too fast ... at any rate not faster than the rate at which the carbon dioxide can be converted by photosynthesis'. He encouraged the development of solar energy, and regarded the eucalypt as the most effective agent for photosynthesis.

'The Keeling curve'

Charles David Keeling (1928–2005), an American chemist, oceanographer, and meteorologist, working at the Scripps Institution of Oceanography from 1956 to 1999, developed an instrument for the accurate measurement of carbon dioxide samples. At Mauna Loa, Hawaii, in the observatory that has been continuously collecting atmospheric data since the 1950s, he measured seasonal variations in carbon dioxide, with a sharp increase overall, known as the 'Keeling curve'—a compelling demonstration of the accumulation of human-produced carbon dioxide in the atmosphere.

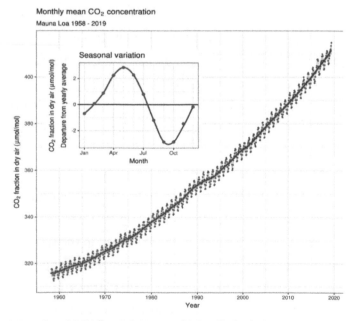

Source: R.F. Keeling, S.J. Walker, S.C. Piper, and A.F. Bollenbach, Scripps CO₂ Program

NASA–GISS and James Hansen

The National Aeronautics and Space Administration (NASA) set up the Goddard Institute for Space Studies (GISS), which American climatologist James Hansen directed from 1981 to 2013. He played an important role in alerting government agencies and the broad population about the need to develop policies to ameliorate climate change. He argued that governments should aim at stabilising carbon dioxide at 350 ppm, while others settled for 450 ppm. For this, he came under fierce attack from the coal industry. Hansen has been involved in a number of lawsuits suing the US government and some of its executive agencies for not protecting a stable climate system.

The National Oceanic and Atmospheric Administration was founded (under another name) in 1807. It monitors the poles (by satellite) and the oceans, collecting and disseminating vast amounts of data. Its budget was cut back during the administrations of George W. Bush and—even more—Donald Trump. Being the US weather service, though, it is still functioning at a reasonable level, and collaborates closely with NASA.

The World Climate Research Program (WCRP)

The WCRP was established in 1979 after the first World Climate Conference, and led to rapid progress in understanding the physical basis of climate change and development of sophisticated numerical models of the climate system.

The Intergovernmental Panel on Climate Change

The IPCC was established by the World Meteorological Organization (WMO) and the UN Environment Program (UNEP) in 1988. In its fourth report (2007), the IPCC concluded that there was a 90 per cent probability that 'the post-industrial rise in greenhouse gases does not stem from natural mechanisms'.

In 2007, the Nobel Peace Prize was shared by the IPCC and Al Gore 'for their efforts to build up and disseminate greater knowledge about man-made climate change, and to lay the foundations for the measures that are needed to counteract such change', a controversial award which was subject to sustained attack.

The IPCC reported in 2019 on attempts to keep global temperature increases down to 1.5°C:

> Many countries considered that a level of global warming close to 2°C would not be safe and, at that time, there was only limited knowledge about the implications of a level of 1.5°C of warming for climate-related risks and in terms of the scale of mitigation ambition and its feasibility. Parties to the Paris Agreement therefore invited the IPCC to assess the impacts of global warming of 1.5°C above pre-industrial levels and the related emissions pathways that would achieve this enhanced global ambition.

In a special report in October 2018 on global warming, the IPCC had estimated that to have a 66 per cent chance of keeping global warming to 1.5°C above the average for 1850–1900 by 2050, the global community would need to adopt a 'carbon dioxide budget' that limited future cumulative emissions to 420 billion tonnes of carbon dioxide (420 gigatonnes of carbon dioxide, abbreviated as 420 GtC).[2]

It warned that exceeding this figure might take the climate system to a tipping point, beyond which changes would be irreversible. Since current emissions exceed 40 GtC per annum, this suggests that the budget will be blown in ten more years. For a 50 per cent chance of limiting global warming to 1.5°C, the carbon dioxide budget increases to 580 GtC—that is, fourteen years of emissions at current rates.

Despite the fierce attacks on its methodology, the conclusions of the IPCC are inevitably conservative, because they are based on consensus and are not exactly new. What is unprecedented is the rapid rate of increase in greenhouse gases and its impact on natural systems. (The IPCC's regular reports can be downloaded free.)

It has to be conceded that the IPCC reports are extremely complex, often qualified, and far from dogmatic. They do not make easy reading, and it is clear that very few of our politicians have read more than a one-page summary of them.

'The anthropocene era'

From 2000, Paul Crutzen and Eugene Stoermer promoted the use of the term 'Anthropocene' to describe a new geological and ecological era:

> To assign a more specific date to the onset of the 'Anthropocene' seems somewhat arbitrary, but we propose the latter part of the 18th century ... we choose this date because, during the past two centuries, the global effects of human activities have become clearly noticeable ... [D]ata retrieved from glacial ice cores show the beginning of a growth in the atmospheric concentrations of several 'greenhouse gases', in particular CO_2 and CH_4. Such a starting date also coincides with James Watt's invention of the steam engine in 1784.[3]

Crutzen, a Dutch atmospheric chemist, shared the Nobel Prize for Chemistry in 1995 with Mario Molina and Sherry Rowland for their work on ozone depletion and the identification of more greenhouse gases.

The science of climate change may be regarded as 'settled', although some aspects are puzzling and will require years of intensive research. But time may not be on the side of the planet, or the next generation. The failure to act effectively by the United States, China, India, Russia, and Australia is essentially political, a triumph of vested interest over the public good.

CHAPTER SEVEN

Climate Change: the politics

> [Non-intervention] is a metaphysical and political word, which
> signifies much the same thing as intervention.
> – CHARLES-MAURICE DE TALLEYRAND (1832)

Climate change poses unprecedented challenges not only to the environment but to democratic practice and the pluralist values associated with Western liberalism. Political, economic, and psychological factors paralyse the will to act to mitigate global warming, leading to denial, prevarication, crude appeals by vested interests, and a growing but sometimes unspoken concern that the climate system may pass beyond a tipping point to become irreversible. Climate change as a *political* issue has probably had a greater effect in Australia than in any other nation. Climate change/global warming has become a central factor in Australia's bitter culture wars. Political responses to it have ranged from paralysis to toxicity.

The role of fossil-fuel lobbyists

Using professional lobbyists to promote scepticism about linkages between fossil-fuel use and climate change was inspired by methods pioneered by the tobacco industry, which dismissed research that identified smoking as a primary cause of lung cancer and tried to prevent government action to curtail it.

Fred Singer (1924–), a retired professor of environmental science at the University of Virginia, and an expert on satellites, was a central figure in both campaigns, and achieved significant media exposure as a 'sceptic' (a description he preferred to 'denier'). He received funding from the tobacco and fossil-fuel industries, but insisted on his independence.

Sir Ronald Aylmer Fisher (1890–1962), an outstanding English statistician, was deeply sceptical about the researches of Sir Richard Doll, Sir Bradford Hill, and others pointing to a correlation between smoking and lung cancer. Fisher insisted that 'correlation' does not mean 'causation', and that other factors—mostly genetic—had to be considered, and denounced the *British Medical Journal* as 'statistically illiterate'. Fisher became a consultant to the tobacco industry—again asserting that the funding did not compromise his research or his conclusion that tobacco was 'a mild and soothing weed'.

The cherry-picking of evidence, and treating anomalies as if they were the norm, became standard practice. It was pointed out that non-smoking nuns died of lung cancer while Jeanne Calment (1875–1997), the oldest person in geriatric records, smoked (but not heavily) until the age of 117.

In Australia, the Lavoisier Group was established in March 2000 by Ray Evans and Hugh Morgan of Western Mining, and proved to be an astonishingly effective lobby, with a powerful network of vested interests and direct access to John Howard and his cabinet.

It campaigned against attempts to secure international agreements to reduce anthropogenic contributions to climate change. Adopting the name of Lavoisier, a great scientist, executed during the French Revolution, demonstrated profound cynicism. Choosing to name an anti-science organisation for a scientific hero was the kind of action that gives cynicism a bad reputation, like naming a gambling syndicate after Gandhi or the Dalai Lama.

I should declare an interest. Ray Evans (1939–2014) was, at one time, a friend, and we worked together on some issues before he defected from the ALP to the Democratic Labor Party. In 1986 he had been a co-founder, with Peter Costello, of the H.R. Nicholls Society, which aimed to destroy the wage-arbitration system; his Bennelong Society (2001) campaigned against native title. He demonstrated how one person, with powerful backing (like B.A. Santamaria, whom he regarded as a model) can transform the political landscape. He argued that climate-change scientists were essentially corrupt, accusing them of 'groupthink' and of seeking huge research grants, while the mining companies were entirely devoted to making life better. The response of the scientists was feeble.

Bitterness is not, I hope, one of my characteristics, but I am prepared to make an exception in the case of Ray Evans.

The American spin doctor and consultant Frank Luntz, author of *Words that Work* (2007), suggested that, rather than attacking scientists directly, global-warming sceptics should use the formula, 'The science is not settled.' His advice was taken up by Australian politicians, from John Howard on, and the formula has been faithfully repeated, year after year.

Luntz argues that '80 per cent of our life is emotion, and only 20 per cent is intellect ... I am much more interested in how you feel than how you think ...'

However, after the 2017 Californian bushfires, Luntz began arguing passionately for action to curb anthropogenic global warming.

Some prominent geologists have been very generous in offering their expert advice about meteorology. Geologists are trained to think about the very long term—millions of years—and a lifetime or two might seem inconsequential to them. Meteorologists have been more restrained, and I can find no examples where they pontificate about geology. Cardinal George Pell, until recently the third-highest official in the Vatican, was a climate-change denialist (unlike Pope Francis), but he never sought environmentalist Tim Flannery's views on theology.

Lobbyists against effective action on climate change assume that since they themselves habitually tell half-truths, exaggerate wildly, advance the cause of vested interests, and engage in character assassination, climate scientists act in the same way.

Some antagonists, including former Liberal prime minister Tony Abbott, denounced the commitment to preserve the environment as a 'green religion', a form of pantheism. Oddly, English environmentalist James Lovelock, having been an early and overzealous advocate for climate-change action, is now retreating—and has endorsed the term 'green religion'. However, Pope Francis is in the other team.

I have come to see the binary distinctions between politics and science on the issues associated with climate change as follows:

Politics	Science
1. Winning the next election	1. Saving the planet
2. Short term	2. Long term
3. Micro	3. Macro
4. Local employment	4. Global priorities
5. Cheaper electricity	5. Reducing climate risks
6. Carbon economy	6. Post carbon economy
7. Generating suspicion	7. Proposing better outcomes
8. Environment as threat	8. Environment as habitat
9. Rejecting complexity	9. Coping with complexity
10. Dismissing expertise	10. Relying on expertise

How good is Australia?

On climate-change questions, not very. The last serious debate in the Australian parliament about any aspect of climate change was in 2009 when the original Carbon Pollution Reduction Scheme (CPRS) was defeated in the Senate. Even then, the primary emphasis was not on the science, but on the economics, politics, and personalities.

In the following decade, the climate-change issue became toxic and infantilised. Neither of the major parties fully understood the science and Australia's role as a world outlier in carbon dioxide emissions.

There has been no discussion in the parliament or the media of SR15 (*Global Warming of 1.5°C*), the 2018 special report by the Intergovernmental Panel on Climate Change (IPCC), proposing a 'carbon dioxide budget'. The report, referred to in the previous chapter, would require cumulative global carbon dioxide emissions between now and 2030 to be less than 420 billion tonnes for the world to have a 66 per cent chance of limiting mean global temperatures rises to 1.5°C.

A number of important questions about climate change have never been asked in the parliament or by the media. Ministers, and especially the prime minister, should be asked:

1. Is it a fact that Australia's mean temperature has already increased by 1.5°C since 1900, long before the rest of the world? The European Union, the United Kingdom, Canada, New Zealand, and the Vatican have already declared a 'climate emergency'. Will Australia, more under threat than most other nations, follow their example?

2. Why has the parliament not debated, or the government not responded to, the IPCC's SR15 report by adopting a

'carbon dioxide budget' to ensure that it contributes to world temperatures not reaching a 'tipping point' of 1.5°C by 2030? Have ministers and party leaders read the report?

3. Why is Australia no. 1 in the world for per capita carbon dioxide emissions (leaving aside oil-producing states such as Qatar, Kuwait, the UAE, and Saudi Arabia), even ahead of the United States, and more than double the European Union average?

4. Is it understood that coal—which accounts for 46 per cent of carbon dioxide emissions—is the primary problem?

5. Is it a fact that each tonne of carbon in coal, when burned, produces 3.67 tonnes of carbon dioxide? (I have never heard this question asked—let alone answered.)

6. Is it understood that carbon dioxide accumulates in the atmosphere, year after year, and that it does not disperse for a century or more?

7. If Australia's coal exports were taken into account (which they are not under the Paris Accords), would our total contribution to world carbon dioxide emissions rise from 1.3 per cent to about 5 per cent?

8. Why does Australia produce three times more carbon dioxide emissions per capita than the United Kingdom and twice as much as New Zealand? Is their quality of life so inferior to ours?

9. Even if we meet our Paris targets 'in a canter', as Prime Minister Scott Morrison keeps on assuring us, will that still leave Australia as no. 1 per capita in carbon dioxide emissions, and are we eager to keep that ranking? If an alcoholic has fifteen standard drinks a day and cuts down to ten, there will be dramatic reduction as a percentage, but she/he is still an alcoholic. The United States ranks as no. 2 per capita in carbon dioxide emissions, but despite President Trump's strong opposition to taking action on climate change, it is cutting back at a faster rate than we are, due to the actions of 41 individual states. California, New

York, Massachusetts, and Oregon each produce about half of Australia's emissions per capita.

10. What is the proportionality of access to Coalition ministers and government MPs from the fossil-fuel lobby (which has a vested interest to promote) compared with scientists in relevant disciplines (committed to the public interest and the long term)? How many ministers, MPs, and parliamentary staffers have professional, family, or other linkages (such as accepting hospitality) with the fossil-fuel industry, especially coal? How many have comparable links with the research community or agencies?

11. Has each Australian state and territory adopted the target of net-zero carbon emissions by 2050, and if so, what is holding the Australian government back from making a similar commitment?

12. Who takes responsibility for cleaning up the mess? We deliver household waste to local councils and arrange for the disposal of old cars, washing machines, televisions, and computers. We are starting to take plastics disposal seriously. Why do we take no responsibility for our fuel wastes? It is an outstanding example of market failure. Is it because—as Tony Abbott thoughtfully suggested—since carbon dioxide is invisible, we don't need to bother?

13. Has the Australian government given any consideration to energy-efficiency measures, so that citizens can retain their quality of life with less waste? (Victoria is offering the free replacement of light bulbs: superior lumens per watt.)

14. Were recent changes weakening the public service's role in providing independent advice to government driven by the climate-change controversy within the aptly named Coalition?

15. Is Australia's take-up of electric cars the lowest in the OECD? If so, why? Tesla's share price is now over $US2,000, and Boris

Johnson is planning to phase out petrol and diesel vehicles. Do Americans and British know something that we don't?

Beyond these vital questions, there is a fundamental requirement for the Coalition and Labor to get serious and cease paying lip service to the science of climate change, and to explain what the central premise is and why it will be necessary to propose unpopular policies to help solve a global problem.

Australia's 2019–20 bushfire season

The Australian bushfire season now arrives earlier and lasts longer. In 2019, it began in late July in New South Wales and intensified in mid-November rather in than January–February; it became more widespread (in four states); and it occurred in tropical rainforests, with conflagrations moving from canopy to canopy, pushed by unprecedented winds.

Then, what had been a long drought was interrupted by record-breaking rainfall, and record heat was followed by record hailstorms.

By definition, surely, this is demonstrable 'climate change', as the Bureau of Meteorology, the CSIRO, university researchers, and state fire-fighting agencies have been arguing for decades.

The Morrison government was open to every explanation for this phenomenon—except one. The bushfire emergency, it contended, could be due to:

- arson (according to Peter Dutton, former Queensland policeman);
- 'lone actors or part of a sinister collective' (Senator Concetta Fierravanti-Wells, former minister)
- policy failures by the ALP and the Greens;
- bad forestry-management practices in limiting the extent of controlled burn-offs;

- children playing with matches (Michael McCormack, deputy prime minister);
- sparks from wires and connections in an overloaded electricity system;
- cigarette embers from passing vehicles;
- exploding horse manure;
- divine punishment for our wicked ways (as in, for example, voting for same-sex marriage) and
- a series of coincidental, but unrelated, incidents.

Or the government provided more comforting explanations such as:

- 'There's nothing new about the current situation: there's been a history of terrible bushfires in Australia from the nineteenth century.' (Scott Morrison); and
- 'Climate is always changing.' (Keith Pitt, the Minister for Resources. He meant 'weather'.)

Even linking climate change with the bushfires was 'political point scoring', and mentioning the issue was a 'knee-jerk' reaction (even after 35 years of concern).

If Australian ecoterrorists were indeed to blame for the bushfires, they must have exceeded their wildest expectations when the Antarctic peninsula reached its highest recorded temperature (20.8°C) in January 2020.

The severity of the Australian fires received extensive international media coverage, and was mentioned in a debate by Democrat candidates for president in the United States. But in Australia there was no serious debate about appropriate action. 'Now is not the time' and 'Hopes and prayers' were the prevailing mantras.

'Climate Emergency'

In 2019, a significant number of parliaments and city councils declared a 'Climate Emergency'. (The City of Darebin, in Greater Melbourne, had been the first jurisdiction to make a declaration, in 2016.)

From late April to late November 2019, 'Climate Emergency' declarations were adopted by parliaments around the world, in this order: Scotland; Wales; the United Kingdom; Ireland; Portugal; Canada; France; Argentina; Spain; Austria; and Bangladesh.

On 28 November, the 28 European Union states (still including the United Kingdom) all adopted the Climate Emergency declaration, and committed to net-zero greenhouse-gas emissions by 2050. This brought in Germany, Italy, Sweden, Norway, Denmark, Finland, the Netherlands, and Belgium.

Pope Francis, the author of the 2015 encyclical *Laudato si'*, about the ethical imperative to take stronger action on climate change, also committed the Vatican to declare a Climate Emergency.

In the United States, a 'Climate Emergency' has been declared by 140 cities, including New York, Los Angeles, San Francisco, Miami, and Austin.

Melbourne, Sydney, Adelaide, and Hobart have all declared a 'Climate Emergency', but no Australian state has, although all are committed to zero-carbon emissions by 2050. It is inconceivable that the current Commonwealth government would make that commitment.

In New Zealand on 8 November 2019, the Climate Change Response (Zero Carbon) Amendment Bill was passed by parliament with only one dissenting vote. The Bill provided for a binding commitment to the Paris agreements, setting the goal of net-zero carbon dioxide emissions by 2050. It was another remarkable achievement by the Labour prime minister, Jacinda Ardern, to persuade the National Party opposition to support the legislation, at a time when Australian politics has proved utterly incapable of reaching consensus on a vital issue for the long term.

By acting unilaterally—and courageously—on greenhouse targets, Australia would not change outcomes on its own, but it could gain moral authority and increase its capacity to play a leading role in international negotiations, encourage many smaller states to commit to strong action, and increase its bargaining capacity with the United States, China, India, and Indonesia.

Why Australia ranks no. 1 in per capita carbon dioxide emissions: the seven 'c' words

Coal, cities, cars, cement, chainsaws, cows, and consumption are the factors that determine Australia's no. 1 ranking of 16.8 tonnes of per capita carbon dioxide emissions. This helps to explain why Australian governments (in contrast, oddly, to the individual states) refuse to take resolute action to reduce the figure.

Australia's electricity generation comes overwhelmingly from fossil fuels. The latest available figures from the Department of Industry, Science, Energy and Resources date from 2018: 60 per cent from coal; 19 per cent, natural gas; and 2 per cent, oil. Renewables provided 19 per cent: 7 per cent from hydro, 6 per cent from wind, and 5 per cent from solar. In 2017–18, about 13 per cent of Australia's electricity was generated off the grid by business and households. Australia's reliance on coal was far higher than New Zealand's or Canada's, but was now less than the United States.[1]

Australia is the world's third-biggest exporter of fossil fuels, behind Saudi Arabia and Russia, and the fifth-biggest miner of fossil fuels (behind China, the US, Russia, and Saudi Arabia). Much of Australia's wealth is based on this, compounding the political difficulty of securing major change.

Australia has a unique tradition of urban development. Sixty-four per cent of the population lives in five cities, the highest proportion in the world, leaving aside city-states such as Hong Kong, Singapore, Vatican City, and the anomalous Uruguay. Just two cities, Sydney and Melbourne, account for 40 per cent of Australia's population.

New Zealand is also highly urbanised, with 54 per cent of its population living in five cities (Auckland, Wellington, Christchurch, Hamilton, and Tauranga).

Then follows Argentina, Bolivia, Greece, Canada, Peru, and Colombia. Eight of the ten most urbanised nations (seven of them in the Southern Hemisphere) are colonial, their urban development being the result, in effect, of an act of state, not of an organic process over millennia, as it had been in Europe. Greece is an exception, essentially because its topography has created enclaves. So is Israel.

The United Kingdom has a population of 68 million, but its five largest cities (London, Birmingham, Manchester, Glasgow, and Newcastle) account for a surprisingly low 24 per cent of the total. It has evolved into a decentralised nation—currently facing the risk of losing its 'united' description—with about 920 towns having a population of 10,000 or more.

All of Australia's major cities have huge energy 'footprints', and impose significant environmental damage. The unique pattern of Australia's urban development has led to a unique environmental impact.

Australia's low-density cities are among the world's largest in area: Melbourne is much bigger than London, Sydney than Delhi, and Brisbane than New York City. Inevitably, this has created a dependence on motor vehicles. The distances between the major cities have also resulted in very long supply chains, reliant on road haulage, trucks, and lorries, rather than on rail (except for transporting minerals to ports).

Australia has 730 motor vehicles (excluding motorbikes) per

1,000 population, ranking no. 10 in the world, behind New Zealand (surprisingly, no. 3, with 860), and the United States (no. 4, with 838).

The rapid growth in Australia's population, especially in large cities, has resulted in the increased use of cement in housing and public works. (Cement can have a negative effect in limiting the soil's capacity to absorb rain, instead channelling it off to the sea.)

Only 16 per cent of Australia's land mass is forested, ahead of the United Kingdom (12 per cent), but far behind Brazil (56 per cent), Canada (49 per cent), Russia (49 per cent), the US (33 per cent), New Zealand (32 per cent), and China (22 per cent). Forests are essential absorbers of carbon dioxide. Australia has one of the world's highest levels of beef production and consumption, at 95 kilograms per capita each year, second only to the US, and ahead of Argentina, quite apart from its major exports of meat, both processed and on the hoof. The methane contribution of what are tactfully described as 'enteric emissions' from beef cattle is significant. (Sheep are a relatively minor factor.)

However, when we examine aggregate carbon dioxide emissions by nations, taking population into account, the ranking is dramatically different:

Aggregate carbon dioxide emissions, as percentages of world total (2017)

China:	27.5 per cent	Germany:	2.0 per cent
United States:	14.7 per cent	Indonesia:	1.6 per cent
European Union:	9.3 per cent	Canada:	1.6 per cent
India:	6.4 per cent	South Korea:	1.5 per cent
Russia:	5.9 per cent	Australia:	1.3 per cent
Japan:	3.0 per cent	Saudi Arabia:	1.3 per cent
Brazil:	2.3 per cent	New Zealand:	0.2 per cent

Australia's figure of 1.3 per cent is the one used by Coalition governments to assert that Australia is not a major contributor to international greenhouse-gas problems, and that there is little we can do (or need to do) to change the global situation.

Climate change and 'revolving door' prime ministers

Controversy about climate change in Australia contributed directly to five prime ministerial removals in eleven years, in a revolving-door phenomenon that became a national speciality.

Climate change has provoked almost unparalleled division and bitterness within the country's political parties, creating a toxic culture in government and leading to policy paralysis.

John Howard was Liberal prime minister when the Kyoto Protocol on Climate Change was negotiated in 1997, setting aspirational (but non-binding) targets for reductions in greenhouse gases. Howard's negotiator, Senator Warwick Parer, the Queeensland Liberal minister for resources and energy, had extensive coal investments.

To offset its production of carbon dioxide from the burning of coal and oil, Australia was able to negotiate significant credits from sharp reductions in land clearing, especially in Queensland, using 1990—a record year for clearing—as a baseline. It was a clever way of avoiding having to take action to cut carbon dioxide production. Despite the free gift to Australia, Howard refused to ratify Kyoto, joining the United States and the micro-states Lichtenstein and Monaco.

Kyoto came into force in 2005 after ratification by 140 signatories, and was operational until 2012.

Howard, Australia's second-longest–serving prime minister, had probably passed his use-by date after eleven years in office (1996–2007). However, his exit was accelerated by Labor leader Kevin Rudd's adroit use of the climate-change issue in the 'Kevin 07' election campaign. The Coalition was defeated in November

2007, and Howard lost his own seat as well. Climate change was not central to his loss, but was a contributing, destabilising factor.

Rudd had declared that climate change was 'the great moral challenge for our generation', with environmental, economic, social, and even national-security implications. He campaigned with conviction on the issue, and when the ALP formed government in December 2007, the first item of cabinet business was the ratifying of the Kyoto Protocol. However, it is sometimes forgotten that Rudd saw Australia's effective action on greenhouse-gas emissions as largely based on 'clean coal'—not just a difficult task, but an oxymoron, like 'non-combustible petrol'.

(We might be able to clean coal a little—but only by a fraction. We take coal's dirtiness as a given when its major advantages are abundance and low cost.)

After Barack Obama was elected US president in November 2008, he indicated support for global action on climate change, but lacked the numbers in the Senate to ratify Kyoto. Canada's Conservative prime minister, Stephen Harper, took his country out of the Kyoto Protocol in 2012.

In Australia, Labor became entranced by the use of the term 'carbon pollution', and avoided references to 'global warming', 'climate change', or 'greenhouse gases', or to explaining the science. 'Carbon pollution' sounded like a clean-air campaign, a public-health measure, of great benefit to asthmatics. Who could possibly object to clean-air measures? But the term 'carbon pollution' was meaningless and unknown to science. Neither carbon nor carbon dioxide are on Australia's National Pollutant Index.

The failure to talk about greenhouse gases or to identify carbon dioxide emissions from coal use as the largest single contributor to climate change/global warming seriously weakened the case for change from the outset.

There was also a lack of candour in refusing to acknowledge

how difficult change would be, since Australia's exports, industry, and electricity supply were overwhelmingly dependent on coal, and early alternatives were not obvious.

The Carbon Pollution Reduction Scheme (CPRS), essentially a 'cap and trade' emissions trading scheme, was twice voted down by the Senate (on 13 August and 2 December 2009): it was rejected by the Coalition (for being too green); by the Greens (for not being green enough); by Nick Xenophon (who preferred an 'intensity-based' ETS, not a cap-and-trade one); and by Steve Fielding of Family First (who didn't understand it).

Malcolm Turnbull, then Liberal leader for the first time, expended a significant amount of his political capital in an attempt to secure Coalition support for its passage through the Senate.

The CPRS model was imperfect, but the Greens made a serious misjudgement in rejecting it—thereby assuring its failure—assuming that a better one would soon turn up. It didn't. (It was another case of 'The perfect is the enemy of the good.')

When the political fix failed, Turnbull lost the Liberal leadership, by a single vote, to a climate-change denier, Tony Abbott, in December 2009. Abbott had famously declared that climate change science was 'absolute crap' and that the world was actually cooling. Labor failed to challenge him on the science.

Once Abbott became opposition leader, the government had to deal with the climate-change issue either by confronting it, by calling a double-dissolution election to establish the CPRS, or by postponing consideration of it. Rudd chose the second option, and lost credibility.

Abbott's claim about cooling related to the observation that 1998 had been (by a fraction) the warmest year on record since 1880. The reason that 1998 looked like a spike on the climate charts was the combination of a severe El Niño and anthropogenic global warming. Over the next few years, temperatures were slightly lower than in 1998, and this became the basis of the assertion, assiduously peddled by shock jocks and climate-change denialists, that the world was cooling. Well, no, it wasn't. The spike in 1998 was an anomaly, and the trend line has steadily gone up ever since. (Strikingly, the year 2019, the warmest on record, had a very weak El Niño. Meteorologists predict that 2020 will be warmer still, setting a new, unwelcome record.)

Also, in December 2009, the Copenhagen Climate Change Conference (known as COP15) failed to secure international agreement to establish binding targets for greenhouse-gas reductions, and was beset by controversy about the validity of projections by some climatologists.

Ironically, COP15 was held during a sudden and exceptionally strong cold snap in the Northern Hemisphere, making it a target for derision by global-warming sceptics. This was profoundly destabilising, and not even President Obama could save it.

Australia had argued that there was a moral imperative to pass climate-change legislation before Copenhagen so that it could play an international leadership role in setting targets. When Copenhagen failed, an uneasy tension developed between the United States and China, and developing nations resisted major changes. The Copenhagen Accord was 'noted', but not 'adopted': it encouraged unspecified action to ensure that global temperature increases be kept below 2°C. Urgency then appeared to drop out of national policy while Australia awaited an international consensus to develop on curbing greenhouse-gas emissions.

Australia retreated from its aspiration to be seen as an international leader and exemplar in setting greenhouse-gas goals. Labor made the inexplicable blunder of failing to involve the community in education and advocacy, despite opinion polls indicating that more than 60 per cent of voters favoured setting a carbon price and acting decisively on climate change.

When Rudd failed to pursue any public involvement in the climate-change debate, and indicated in April 2010 that action on the issue would be postponed until 2013, incredulity and disillusion were immediate. The sudden fall in Rudd's approval ratings began a process that led to his removal from office in June 2010 by factions dominating the ALP caucus. Julia Gillard won the leadership unopposed, after Rudd declined to contest the ballot and resigned.

In the August 2010 election, the ALP was rather defensive on climate change, and Gillard's promise that 'There will be no carbon tax under the government I lead' was injudicious and probably unnecessary.

After the election, her government needed the support of independent MPs Tony Windsor and Rob Oakeshott to survive. They wanted a carbon-pricing mechanism, which was duly announced in February 2011. This was not a tax, strictly defined, but an inducement to persuade citizens to change their patterns of consumption.

Abbott then launched savage, relentless, and misogynistic attacks on Gillard. Labor was feeble in response, exposing her to unprecedented damage, compounded by public protests about the 'carbon tax' and the combined weight of the shock jocks and News Corp papers.

But Abbott directly addressed *voters*. Whereas the government retreated into silence about a complex, badly designed, and woefully explained scheme in an area that was contested and poorly understood, Abbott offered a six-word response ('A great big tax on everything') directed at voters' fears, and it was highly successful. The government had acted courageously in setting a carbon price of $23 per tonne (rather high, as it turned out), and offered compensation to consumers on low incomes, but then failed spectacularly to explain why it was being done.

In July 2011, the Gillard government circulated to every

household in Australia a twenty-page booklet (including eighteen pages of text) entitled *What a carbon price means for you,* designed 'to help you learn more about the financial assistance you and your family may receive'.

There was not one sentence about climate change, or an explanation of why carbon pricing was being introduced. It was the weakest argument for an important national policy change since the official 'Yes' case in the referendum for an Australian republic in 1999.

Rudd returned as prime minister in June 2013, but his government was comprehensively defeated in September, and Abbott formed a Coalition government. In 2014, the carbon-pricing legislation was repealed by his government.

Abbott had been a deadly opposition leader, but as prime minister he failed, making it clear that protesting and attacking, not governing, was his metier.

Malcolm Turnbull defeated Abbott in a Liberal party-room ballot in September 2015, was sworn in as prime minister, and attempted to produce a policy on climate change and energy that would not split the Coalition. Despite being the recipient of enormous goodwill when he replaced Abbott, Turnbull won the 2016 election with a bare one-seat majority, which weakened his authority, making his survival dependent on the National Party—a party dominated by climate-change sceptics and deniers.

In February 2017, in the House of Representatives at Question Time, Scott Morrison, then treasurer, answered a so-called Dorothy Dix question from a Coalition backbencher. He denounced the Labor opposition for its 'pathological, ideological fear of coal', and waved around a lump of coal to emphasise how essential coal was for economic prosperity and jobs, especially in the regions. (Turnbull kept his head down, and did not look at Morrison, during this performance.)

Turnbull's National Energy Guarantee (with the unfortunate acronym NEG) attempted to ensure energy security while retaining a commitment to reduce carbon dioxide emissions. It was cautiously welcomed by business as providing some sense of direction, after years of drift. But when Turnbull failed to get support from Abbott and his close allies, he abandoned the NEG. Within a few days, the Liberals abandoned him, and Morrison became prime minister in August 2018.

'The Australian people have spoken.'

Public opinion polls indicate high levels of support for stronger action on climate change, surprisingly so, given that the science has been so abysmally discussed. After the May 2019 election, which the Coalition won narrowly, Michael McCormack, the deputy prime minister and leader of the National Party, intoned that, on climate change, 'The Australian people have spoken.' Indeed. But what did they say?

The ALP, which had an imperfect policy on tackling climate change, but one that was far stronger than the Coalition's, won a majority of seats in New South Wales and Victoria, half the seats in South Australia, and all the seats in the Australian Capital Territory and the Northern Territory. With Andrew Wilkie, progressives won three of Tasmania's five seats.

On climate change, Labor opposition leader Bill Shorten had proposed significant increases in emissions reductions without explaining adequately why a Labor government would do so. Labor's climate-change policy was poorly argued, failing to involve millions in the community who engage directly with the issue (such as gardeners, farmers, bushwalkers, anglers, bird watchers, whale watchers, beekeepers, and skiers). Their lived experience and direct observation should have been harnessed, but was not.

Shorten never talked about the science, and he rejected my advice (and maybe others') to emphasise the fact that each tonne of coal burnt produces 3.67 tonnes of carbon dioxide, which hangs round for decades in the atmosphere as a greenhouse gas. It is not 'demonising' coal to point this out. And he never referred to cement, cows, and cities, all three presenting central difficulties in reducing greenhouse-gas emissions. He talked about electric cars, but not persuasively. And he invited criticism that he was straddling the fence on the proposed Adani coal mine in Queensland, not wanting to offend the Construction, Forestry, Maritime, Mining and Energy Union (CFMMEU), which had provided him with critical factional support.

Labor lost two seats in Queensland, and failed to win any in Western Australia—both mining states—but the largest gains in primary votes were achieved by far-right parties. The Coalition's vote dropped by 3.5 per cent in Western Australia, and rose by only 0.5 per cent in Queensland.

The Greens polled many more primary votes than the Nationals (including Liberal National Party MPs from Queensland, who caucus with the Nationals). The Greens secured 1,482,883 votes, 10.4 per cent of the national aggregate, but won only one seat in the House of Representatives. They won six Senate seats (holding a total of nine).

The National Party of Australia, with 914,246 votes, 6.4 per cent of the aggregate vote, won sixteen seats in the House of Representatives, and has five senators. In the Coalition, the Nationals provide the deputy prime minister and six ministers.

Senator Matt Canavan, a Queensland member of the Liberal National Party, and a former minister for national resources, talks like a candidate for shire president: all his priorities are obsessively local. Collinsville, in central Queensland, has a population of 1,500, and two coal mines operate there. Controversy has erupted about the Commonwealth's funding of a feasibility study, costing $4 million, for a third mine. Oddly, this is not a feasibility study for the Commonwealth, but for the mine's proponents—essentially a free gift.

I assume that if Senator Canavan was asked his priorities, whether to attempt to save the planet or create more coal-related jobs in Collinsville, he would choose Collinsville. If I were a National Party senator from Queensland, with aspirations for leadership and a brother in the coal industry, I might make the same choice—but I'd hope not.

In her powerful *Quarterly Essay* 'The Coal Curse', Professor Judith Brett observed that, 'The National Party has become the party of coal', that its former leaders Mark Vaile and John Anderson 'made fortunes out of resources', and that when Waleed Aly challenged Michael McCormack (March 2019) to name a single case where the Nationals 'had sided with the interests of farmers over ... miners', he failed to do so.

(Readers might note an eerie similarity between the case that Judith Brett and I are arguing, and the examples given, but I was already correcting my proofs when the *Quarterly Essay* appeared. Clearly, we think alike.)

In 2019, the highest swing against an ALP candidate anywhere (14.2 per cent) was in Labor MP Joel Fitzgibbon's mining seat of Hunter in New South Wales. The One Nation candidate, a miner and member of the CFMMEU, recorded a primary vote of 21.6 per cent. Fitzgibbon, Labor's mirror image of Canavan, blamed Shorten for not supporting coal, but I suspect that much of Fitzgibbon's bad polling was all his own work.

Much fear and loathing is created by exploiting the anxiety of workers in economic monocultures—such as coal-mining, forestry, sugar production, electricity generation, aluminium-smelting—and telling them, 'Only one type of employment is possible for you and your family.'

But it is essential to talk directly to the coal miners and their families—as Bob Hawke did to tobacco, motor-manufacturing, and clothing workers in the 1980s—and say, 'You won't like what I am about to tell you, but you have to face reality. Sudden change will be very tough for you, but we have to think ahead for the next generation. How many of you hope that your children will be working in coal mines or tobacco? Would you prefer them to have a range of choices? And how can we help now?'

There are about 37,000 coal miners in Australia (0.3 per cent of the labour force)—and some 200,000 other workers (1.5 per cent) indirectly involved.

Present Labor leader Anthony Albanese has committed the ALP to the net-zero carbon emissions target, which I applaud, but has set no priorities, has never talked about the science, and his lips have never uttered the words, 'Coal is the problem.' In fact, he sees a future for the coal industry, but says that the market should determine if a project is viable, with no government investment.

Mark Butler, Labor's shadow minister for climate change and energy, has been a strong advocate for reducing emissions. But, mysteriously, in his book *Climate Wars* (2017) he devotes only a page and half to the science that is the prime justification for taking action on climate change.

In Queensland, under the Labor government of Annastacia Palaszczuk, the projected Adani Carmichael coal mine stands to benefit from $4.4 billion over 30 years in royalties foregone, free water, tax concessions, and subsidies.

Lobbyists and parliamentary staffers (often known as 'minders') have become an important—and highly secretive—element in modern government, and the transition from staffer to MP to minister to lobbyist is a critical element in how and why decisions are made.

The role of 'minders' has been exempt from parliamentary scrutiny, such as Senate select committees. As it happens, the minerals lobby, especially 'big coal', has been an important factor in shaping policy on climate change and giving a sharp nudge to pushing prime ministers down a mine shaft.

In the Coalition government, the links with fossil fuel and mining are particularly strong, especially in Scott Morrison's office and the National Party:

- Sir Lynton Crosby and Mark Textor run the powerful lobbying and political strategy firm Crosby Textor, whose clients include Glencore, Mitsubishi, and the Queensland Resources Council. Crosby was federal director of the Liberal Party 1997–2002.
- Yaron Finkelstein, Scott Morrison's principal private secretary, Andrew Hirst, the Liberal Party's campaign director, and Queensland Liberal National Party senator James McGrath all worked for Crosby Textor.
- Dr John Kunkel, Morrison's chief of staff, worked for Rio Tinto.
- Ian Macfarlane retired as minister for industry to become CEO of the Queensland Resources Council.
- Angus Taylor, the Liberal minister for energy and emissions reduction, was a consultant for the Minerals Council of Australia (MCA).
- Melissa Price, the Liberal minister for the environment 2018–19, who was kept out of sight during the 2019 election campaign in case she was asked questions, was vice president of legal and business development for Crosslands Resources, a mining company owned by Mitsubishi.
- Barnaby Joyce, a former National Party deputy prime minister, has been described as a wholly owned subsidiary of Gina Rinehart of Hancock Prospecting.

- Liberal National Party senator Matt Canavan was minister for national resources. His brother, John, worked for global coal-company giant Peabody Energy, and is a part-owner of the Rolleston coal mine.
- Sophie Mirabella, a former Liberal MP, now works for mining magnate Gina Rinehart.

Is there significant exposure to scientific expertise inside ministers' offices? The chief scientist, Dr Alan Finkel, plays an important role, but I assume he is at arm's length from the prime minister's office.

Leadership

Leadership demands courage, long-sightedness, hard knowledge and, often, the preparedness to tell voters what they don't want to hear.

In 1982–83, Labor committed itself to oppose the flooding of the Gordon-below-Franklin wilderness area in south-west Tasmania, a World Heritage site, arguing that its preservation was more important than creating another dam to produce hydroelectricity. This was unpopular in Tasmania: in the 1983 federal election, the ALP primary vote slumped across the state, and no seats were gained. However, Bob Hawke led Labor to a comfortable victory with major gains on the mainland. Labor gained credibility because of its commitment to a principle.

If a comparable challenge arose in Tasmania in 2021, I doubt that the ALP would take the same courageous and principled stand. I suspect that the mantra 'Jobs! Jobs! Jobs!' would win out.

The Gordon-below-Franklin wilderness was saved, and is no longer an issue in the state. By 1989 Tasmania had an ALP state government, and by 1993 Labor held four of the state's five electorates in the House of Representatives.

When in government, federal Labor had been courageous in adopting a carbon price, but failed to explain it. Since its defeat in 2013, the ALP recognises that the problem of climate change is huge, but the remedies it has proposed have been tiny, determinedly anti-Churchillian, painless, and without risk, sacrifice, or the need to change behaviour or patterns of consumption. This lacks conviction or psychological carrying power.

It is implausible to argue that a fundamental change in national priorities and economic base would have no impact on employment or consumption.

Just imagine the same stance being adopted with significant social problems: 'We will reduce juvenile obesity—and subsidise junk-food outlets as well, to ensure no job losses. We'll reduce gambling, but at no cost to Crown Casino. We'll cut binge drinking, but not at the expense of manufacturers or retailers. We'll reduce road trauma, but with no job losses in the vehicle-repair industry. We'll be tough on smoking, but no tobacconist will have to close.'

Strikes by school children in Australia during 2019 over climate inaction, initiated in Sweden by the young activist Greta Thunberg, caused particular irritation to News Corp newspapers and Coalition politicians. The largest strike—100,000 strong—was held in Melbourne on 20 September 2019.

Scott Morrison remarked, with his usual acuity, 'Let kids be kids', which was coded language for, *Why don't they shut up about climate change, because we are so much better informed on the issue?*

Hattie O'Brien, in the *New Statesman* (20–26 September 2019), linked climate denialism to an aggressive masculinity that resents challenges to the way the world is exploited. Canada's environment minister, Catherine McKenna, said, 'Misogyny and climate denial seem to go together.' Tony Abbott's attacks on Julia Gillard, Alan Jones's on Jacinda Ardern, Donald Trump's on Angela Merkel, and Jeremy Clarkson's on Greta Thunberg all had a manic quality.

Morrison's claim that Thunberg was provoking 'needless anxiety' among children was demeaning and condescending. Dr Jennifer Marohasy of the Institute of Public Affairs is the only prominent female climate-change sceptic—other than Gina Rinehart—whom I can think of. This lends weight to the misogyny hypothesis.

The Business Council of Australia, the National Farmers Federation, and insurance companies want clear policies on climate change, and banks no longer lend to expand coal production. (It is surprising that major superannuation funds, including AustralianSuper, Cbus, Hesta, and Unisuper, still retain fossil-fuel investments. However, individual superannuants are able to opt out.)

Business leaders who called for stronger action were criticised for exceeding their brief and pursuing any issue other than shareholder returns. They included Andrew Mackenzie, FRS, the outgoing CEO of BHP, a distinguished scientist himself, and Alan Joyce, CEO of Qantas, who was eager to cut aviation emissions. Leading global mining group Rio Tinto is working towards net-zero carbon outcomes.

Climate-change activism was often characterised by politicians as an attack on consumption patterns and the idea of growth as an end in itself.

Risk management: 'the 1 per cent doctrine'

Oddly, the sceptics/denialists never discuss the question of 'risk'. And yet the key element in climate-change mitigation is risk management in an uncertain environment.

Dick Cheney was the formidable vice president of the United States from 2001 to 2009, under George W. Bush. In November 2001 he proposed 'the 1 per cent doctrine'. This was the argument (essentially about terrorism, after the 9/11 attack in New York) that, 'If there is a 1 per cent chance that a threat is real, we have to treat it as a certainty in terms of our response … It's not about our analysis or finding a preponderance of evidence.'

Ironically, for climate-change denialists, even a 90 per cent probability is not enough to change their mindset.

Clean air, clean water, the preservation of species—including bees, insects, and birds, with their critical role in agriculture, and fish, and biodiversity generally—are not left or right issues. All are long-term rather than short-term matters.

Climate-change denialists or sceptics fail to offer an alternative hypothesis to explain the observed phenomena. They simply deny that global warming is occurring, asserting (again, without evidence) the case that the atmosphere is cooling and that scientists involved in the IPCC are engaged in self-promoting fraud, while lobbyists for the fossil-fuel industry operate only from the purest motives.

Our priorities must be to start with the evidence, the data, and the science that interprets it, and then to take what action we can to save Planet Earth, our only home, because actions taken now will have a cumulative effect, well beyond our lifetimes. Psychologically, it is virtually impossible for most of us to imagine what the Earth will be like in the year 2100. All we know is that very few of our contemporaries will survive to that date. But our children and grandchildren will.

So we must think of future generations, and consider ethics ('Is it right?') from a longer-term perspective.

But our habitat—clean air, clean soil, clean rivers and oceans— must be preserved for the present generation. This involves tackling very complex and competing national and regional issues.

It may seem naive to hope, but if we start with public interest, the likeliest outcome will be to advance individual benefit as well. If we reverse the two, both self-interest and public interest will suffer.

Between now and 2050, when the consensus of expert opinion is that the world must achieve net-zero emissions of greenhouse gases, there will be ten elections for the Commonwealth parliament.

Every minister in the Morrison government would say: '2050 is a long way off. Cutting to net-zero emissions can be left to the government in power in the 2040s'.

In 1992, I proposed the following variant of 'Pascal's wager'[2] to illustrate the options for climate change:

- If we take action on climate change and disaster is averted, there will be massive avoidance of human suffering.
- If we take action and the climate-change problem abates for other reasons, little is lost and we benefit from a cleaner environment.
- If we fail to act and disaster results, then massive suffering will have been aggravated by stupidity.
- If we fail to act and there is no disaster, the outcome will be due to luck alone, like an idiot winning the lottery.

Angus Taylor asserts, 'Australia cannot act unilaterally' in accepting a zero-carbon target. But Australia *is* acting unilaterally.

Seventy nations have now adopted the goal, and Canada, the nation most closely resembling Australia economically, has pledged itself to an 80 per cent reduction in carbon dioxide emissions. Australia allies itself with the outliers—China, the United States, and India—but (to repeat) our per capita carbon dioxide emissions remain the highest of all developed nations.

Scott Morrison has a weird idea that the future is simply a linear projection of the past, so that if your father was a coal miner or a policeman, that's what you will be—and your sons, too. In his vision, if you want to communicate but are away from home, you look for a public telephone or post a letter. If you want some money, you cash a cheque. It's not back to the 1980s, more like the 1950s.

Lived experience confirms that change is usually rapid, unpredictable, and transformative.

Failure to act appears to favour the present, but it certainly prejudices the future. Political leaders find it difficult to recognise a new paradigm.

The astute French diplomat and minister Charles-Maurice de Talleyrand observed in 1832 of non-intervention: '*C'est un mot métaphysique, et politique, qui signifie à peu près la même chose qu'intervention*.' The original is subtler, and I have included the literal translation as the epigraph to this chapter, but I paraphrase it as, 'Not to choose is to choose.'

Not acting to mitigate climate change is not neutral: it accelerates the challenge.

As our charismatic deputy prime minister, Michael McCormack, said, if we increase Australia's emissions reduction target to 45 per cent, then 'Forget night footy, forget night cricket, and you'll have pensioners ... shivering all winter, and ... melting all summer'. Can he be serious? If Australians face the alternative of either making a serious contribution to preventing global temperatures reaching a tipping point or preserving night cricket, are we so shallow and unthinking as to pursue the soft option? If the British and Irish can do it, why can't we?

CHAPTER EIGHT

Retail Politics: targeted, toxic, trivial, and disengaged

The central question in contemporary politics is not 'Is it right?' but 'Will it sell?' 'Retail politics', sometimes called 'transactional politics', where policies are adopted not because they are right but because they can be sold, is a dangerous development and should be rejected. We must remain confident that major problems can be addressed—and act accordingly. This involves reviving the process of dialogue, and rejecting mere sloganeering and populism. We need evidence-based policies, but often evidence lacks the psychological carrying-power generated by appeals to prejudice or fear of disadvantage. ('They are robbing you ...')

In Australian elections for the House of Representatives, where the fate of governments is determined, most electors cast an 'instrumental' vote in what is essentially a two-horse race between the Coalition (Liberals + Nationals) and the Australian Labor Party (ALP).

These are the country's hegemonic parties, essentially mirror images of each other. I lump the Liberals and Nationals together, in practice, despite their often-significant differences on ideology, personalities, and entitlements to the spoils of office. They emphasise individual effort and rewards, but are eager to offer collective help such as fuel and water subsidies.

Labor, historically, has emphasised the collective, as with Medicare and compulsory superannuation.

Both hegemonic parties have become 'shelf companies', with small, ageing memberships, large numbers of 'stacks', and are oligarchic, not democratic, in practice. Party activists are, in effect, traders, apparatchiks, and future lobbyists. They are characterised by personality cults, factions, and ethnic recruitment.

The parties are protected by measures that were originally intended as reforms: compulsory voting; and the public funding of election campaigns (receiving $2.62 for each vote they garnered in the previous election).

I support these policies in principle, as part of Australia's electoral system—probably the world's best, which we should be proud of. But there is a dark side to compulsory voting and the public funding of elections: both virtually armour plate the hegemonic parties.

Parties do not have to interact with the electorate to bring out the vote: it comes out anyway, by force of law. They do not need a significant membership to interact with local communities. Party members could be on life-support systems—or dead (and both propositions may well be true).

Both parties begin each campaign with a large treasure chest, but they always want more, so campaigns are augmented by donations (a.k.a. bribes)—in the Coalition's case, from major interest groups such as gambling, retailing, and mining, and in Labor's case, from the trade union movement.

After the Morrison government's unexpected re-election on 18 May 2019, the first item on cabinet's agenda would likely have been, 'What do we need to do to win in 2022'. Setting long-term objectives would have been dismissed out of hand.

And there were debts to be repaid, particularly to the Clive Palmer-UAP (United Australia Party), whose advertisements, plastered across the wide, brown land, charged that Labor would impose 'trillions' of dollars in extra taxation, and generated fear, especially among older voters, that their thrift would be punished. This was essentially a free $83 million gift by Palmer to Morrison.

Nixon's 'silent majority'

The politics of the United States, and of the United Kingdom and Australia, too, have been deeply influenced by a political strategy devised in 1971 by the White House publicist and strategist Pat Buchanan, and adopted by President Richard Nixon, designed to maintain the status quo and preserve conservative hegemony.

Richard Milhous Nixon (1913–1994), the 37th president of the United States between 1969 and 1974, was the only one to resign his office, when he faced the near certainty of being impeached by the House of Representatives and removed by the Senate over the Watergate scandal.

He had been relatively progressive on some issues, such as promoting legislation on clean air, clean water, and endangered species; creating the Environmental Protection Agency; and expanding Medicare. He accepted the idea that governments could intervene to correct market failures: as he said, 'We are all Keynesians now.' Although he made his early reputation as a fervent anti-communist, his famous visit to Mao Zedong in Beijing in February 1972 ended much Cold War hysteria about the People's Republic of China.

The central feature of the Nixon strategy was to recognise existing divisions in society, identify a sense of grievance or alienation, and exploit them to win elections by, in effect, cutting a country like a cake and taking the bigger slice: white English-speaking voters with an explicitly—or even vaguely—Christian or Jewish background.

The Nixon strategy had several components:

- Appealing to 'the silent majority', drawing support from citizens who were politically disengaged, and who had never spoken out or protested in any way about issues such as segregation, women's rights, the Vietnam War, the death penalty, or the environment. In effect, activism was characterised as bad-mannered, divisive, and even unpatriotic, while silence was applauded.
- Implementing the 'Southern Strategy', a successful exploitation of reactions against Lyndon Johnson's *Voting Rights Act* (1965)— which attempted to ensure that African-Americans were entitled to vote in every state—combined with a dog-whistling appeal to racism. This resulted in a political transformation in the southern states and the old Confederacy. Overwhelming support for the Democrats, which had won the South consistently since the Civil War ended, was converted to Republican domination.
- Identifying and wedging divisions between the Old Left, which was socially conservative on some issues, including sexuality and immigration, and the New Left, with its emphasis on environment, gender issues, race, Indigenous peoples, refugees, law reform, and the arts.
- Promoting issues, such as abortion, state aid for church schools, same-sex marriage, and the ability to hire and fire on religious grounds, that would activate many Catholics and most members of Protestant evangelical churches.
- Promoting hostility to experts, characterised as 'elitists', remote from the lives of ordinary citizens, so that important findings on climate change, health, and diet, and international comparisons on longevity, education levels, and consumption patterns could be attacked or dismissed.

• Reviving the concept of 'political correctness'. Ironically, this was originally a Stalinist coinage by the Communist Party to enforce the party line. It has been adopted by the right, asserting that attempts to control language, driven by 'elites', imposes limitations on speech, belief, or thought—for example, by censuring exclusionary or derogatory language against minorities.

Although he was forced from office, the Nixon strategy has long survived him. In Australia, John Howard was a dog-whistling virtuoso who enthusiastically adopted the Nixon strategy.

Race was, and remains, the most divisive factor in United States politics, is important in the United Kingdom, and is central to 'the great Australian silence'. And religion comes next.

Nixon's 'silent majority' had an important Australian precedent: the appeal to 'the forgotten people', a coinage by Robert Gordon Menzies (1894–1978), Australia's longest-serving prime minister and the founder (1944) of the Liberal Party of Australia.

Menzies, an outstanding Melbourne barrister, had broken periods as prime minister, in 1939–41, when he led the fractured United Australia Party (UAP—a name later purloined by Clive Palmer), and in 1949–66, when he bestrode Australian politics like a colossus as leader of a party he had created, the Liberals, always in coalition with the Country Party (later renamed National Party).

In weekly broadcasts during his wilderness years, published as *The Forgotten People* (1942), Menzies appealed to 'the real life of this nation', to be found not in the 'so-called fashionable suburbs, or in the officialdom of the organised masses', but among the 'nameless and unadvertised' people whose homes were 'the foundation of sanity and sobriety' and 'the indispensable condition of continuity'.

Menzies' electoral base was essentially in the suburbs. He won seven elections straight—1949, 1951, 1954, 1955, 1958, 1961, and 1963—admittedly helped by a disastrous split in the Australian

Labor Party over attitudes to communism during the Cold War.

Sir Robert retired as prime minister in 1966, two years before Scott Morrison was born. He came to disdain Harold Holt, John Gorton, Billy McMahon, and Billy Snedden, the Liberal leaders who succeeded him. (Malcolm Fraser was the exception.)

John Howard saw Menzies as an inspiring model (they met only once, briefly), and Morrison invokes the names of both.

John Howard's legacy

John Howard's four election triumphs (in 1996, 1998, 2002, and 2004) and his long term as prime minister (1996–2007) represented a fundamental change in Australian political history and a strong reaction against the reforms of the Hawke–Keating years (1983–96), in an unprecedentedly successful Labor government.

Although we do not appear to have much in common, and I disliked many of his policies, I always had an amiable relationship with John Howard. At least until now.

I admired, reluctantly, his exceptional gifts as a tactician, and, as conceded earlier, his stand on guns, tobacco and—to a degree—the GST. His memory for people is unsurpassed in my experience. (Bill Clinton had the same gift, plus charisma.)

In Howard's first year as prime minister, seven of his ministers resigned over conflicts of interest or dubious claims for travel or other allowances. Many of my parliamentary Labor colleagues saw this as a sign of weakness, and predicted that his government would not last long. I saw it as a sign of strength, and he held on for another decade. Ethically challenged Liberal minister Angus Taylor would not have survived in a Howard government, and the Coalition's 'sports rorts' scandal would have been swiftly sorted. This is one part of the Howard legacy that has been abandoned, regrettably.

I disliked intensely his stance on settler history; native title;

refugees; the Norwegian freighter M.V. *Tampa* carrying asylum seekers; 'children overboard'; climate change; the Republic; the Iraq war; politicising the public service, philistinism in the arts; and allowing covert racism to return to the political agenda after 30 years in which both parties had eschewed it.

In 1996, on the ABC's *Four Corners* program, when asked by Liz Jackson about his vision for the year 2000, he replied that he would like to see Australians 'comfortable and relaxed about three things … their history, the present and the future'. He emphasised the role of small business, saw Australia as 'a unique strategic intersection between Europe, North America, and Asia', and downplayed the role of 'excitement'. This was code for repudiating the 'big picture' thrills-and-spills approach of the Hawke–Keating years, whose achievements still largely survive.

Between 1947, when Labor began Australia's mass-migration program, and 1996, when Howard became prime minister, Australia had a bipartisan policy on immigration and refugees. Over 50 years, no major party leader (and there were sixteen of them) ever played the race card until John Howard's two periods leading the Liberal Party. It paid off for him handsomely.

After 1996, racism became once more an important issue in Australian politics. Howard was a cultural warrior, deeply opposed to the pejoratively named 'black armband' narrative of Australian history, committed to the values of a settler society, and willing to encourage white fears that the recognition of native title after the *Mabo* and *Wik* decisions of the High Court would drive Australians of European descent from their homes and land. He co-opted many of the policies articulated, in a shriller fashion, by Pauline Hanson and her One Nation Party and won the support of many former Labor voters, known as 'Howard's battlers'.

He was the archetype of White Australia, the epitome of settler values, the apostle of *terra nullius*.

In September 1997, on ABC television, in a direct appeal to fear and ignorance, he held up a map of Australia, claiming that, if the native title act was not amended, Indigenous Australians would have 'the potential right of veto over further development' of 78 per cent of the landmass. (By 2015 only 32 per cent of Australia had become subject to native title, almost all in very remote areas.)

On 8 October 2001, just after the campaign for the November election began, the government claimed that asylum seekers had attempted to throw their children overboard to force the Royal Australian Navy to take them to Australia, and produced photographs that seemed to support the claim. (Naval personnel disagreed.) This confirmed John Howard's conviction that we did not want 'those kinds of people' in Australia. The claims about 'children overboard' were subsequently discredited completely.

On 19 October 2001, 353 boat people drowned when a vessel code-named SIEV X [Suspected Illegal Entry Vessel—X 'unknown'] sank inside international waters, south of Sumatra, at a time of intense surveillance to our north, without any attempt to save them.

Tampa, 9/11, 'children overboard' and even SIEV X proved to be bankable political assets for Howard, especially when the Labor opposition, fearful of a patriotic backlash, chose to adopt an ultra-cautious 'small target' policy and avoided serious questioning of the government. It was not Labor's finest hour.

Howard took pragmatism to extremes, but lacked intellectual curiosity. He took a restricted—even constricted—view of Australia's capacity to innovate, which could be summarised as *It cannot be done, it is beyond us, and if we tried, we'd muck it up*. In 2007, he was defeated by Kevin Rudd and lost his own seat.

Howard rejected Rudd's 'National Apology to the Stolen Generations' in 2008, arguing that it imposed a burden of guilt (or shame?—he conflates the two) on white settlers (oppressors?) and their descendants. This proved to be central to his post-prime ministerial appeal, and he retains iconic status among the Liberal faithful.

In July 2011, Anders Behring Breivik, a Norwegian far-right terrorist, fanatically anti-Muslim and anti-immigrant, massacred eight people in Oslo, and 69 on the island of Utøya, wounding 319 more. All of his victims were Europeans. In his 1,500-page manifesto, *2083: a European declaration of independence* (2011), he singled out five Australian conservatives for praise: John Howard, Peter Costello (Howard's treasurer), Cardinal George Pell, historian Keith Windschuttle, and former Liberal MP and later talk-show host Ross Cameron.

Morrison's 'quiet Australians'

Nixon's 'silent majority' has its exact counterpart in Scott Morrison's 'quiet Australians', who may never have advocated mistreatment of Indigenous Australians, or supported racism or misogyny or wage theft, but had never said or done anything to oppose them either.

By 'quiet', he means 'passive' or 'docile'.

Morrison delivered a speech to the Institute of Public Affairs on 18 August 2018, warning public servants not to tell ministers what they didn't want to hear and instead to concentrate on the needs of 'quiet Australians'. He wanted officials to ignore people 'in the Canberra bubble' who might be activists. (He did not mention highly paid lobbyists in the same town.)

The Coalition strategy is based on the assumption that securing additional rights for one group necessarily involves a loss of rights for others, creating a resentment that can be harvested in votes.

Recent examples have included claims that:

• professional expertise and evidence restricts the expression of deeply held contrary community opinions (such as on climate change, vaccination, and fluoridation);

- preserving water flows for the environment challenges the viability of irrigators;
- planning for a post-carbon economy threatens mining jobs;
- tackling climate change might drive up power bills for 'hard-working Australian families';
- preserving wilderness values in National Parks limits the rights of bikies to use the tracks;
- securing equal rights for women infringes male hegemony, and tolerating Muslims or LGBTIs imposes a vow of silence on fundamentalist Christians or straights;
- curbing abuse at sporting matches infringes basic sledging rights; and
- restricting trolling and humiliation on social media limits free speech.

These and similar examples fuel constant, but often incoherent, attacks on 'political correctness'. It is deeply felt, but poorly articulated.

In Australia, opposition to 'political correctness' is coded language for the support of White Australia, settler values, Australia Day, and the Australian flag; the celebration of Gallipoli and the monarchy; and uneasiness about multiculturalism. The right now uses the term to deride the use of gender-neutral terminology ('chair' for 'chairman') and the restoring of Indigenous names ('Uluru' for 'Ayers Rock'), and to justify the use of discriminatory language ('black bastard', 'ape', 'coon', 'King Kong', 'poofter', and 'lesbo') as robust examples of free speech.

I suspect that many Liberal voters feel uneasy about discussing politics, whereas Labor voters can't help themselves. While I have followed a strict rule all my life of never asking a person how she votes—after all, we have a secret ballot—when someone says, 'I don't take much interest in politics', it is reasonable to assume that they are committed to the status quo.

My father (not a Labor voter) used to say that Liberal voters could be identified because they always shined their shoes.

An authoritarian leader in Australia seems implausible, largely due to the constraints of the parliamentary system. But the authoritarian style has its supporters. Morrison's style could be called 'soft authoritarian': he bristles when questioned or challenged, or when the 'daggy dad' nice-guy persona does not work.

In the Australian federal election held on 18 May 2019, of 14,250,636 formal votes, 1,300,400 (9.1 per cent) were cast for parties with a far-right, nativist, populist, isolationist, religious, or nationalist agenda—although a lack of policy clarity, especially in the case of Clive Palmer's United Australia Party, creates some problems in classification. Almost all rejected neoclassical economics.

Senator Cory Bernardi's Australian Conservatives disappeared without trace. David Leyonhjelm's Liberal Democrats were also eliminated: progressive on some social issues, such as drugs and marriage equality, they were against tougher gun-control laws, and hard to place on the spectrum.

Pauline Hanson's One Nation Party secured 5.4 per cent of the nationwide vote, retaining one Senate seat. The Katter Australian Party (KAP), not always conservative on some social issues, returned Bob Katter, overwhelmingly, in his electorate of Kennedy. Clive Palmer's UAP spent more than any other party, and won a national vote of 3.4 per cent, but no seats. Fraser Anning's Conservative National Party gained 0.54 per cent of the national vote.

The alt-right voting bloc provided the winning edge for Scott Morrison and the Coalition, which will be increasingly dependent on it in future elections.

In Queensland, of 2,829,033 formal votes, 469,377 (16.7 per cent) were cast for far-right parties, and in eleven seats north of Brisbane the proportion ranged from 20.1 per cent (Groom) to 49.9 per cent (Kennedy).

Religion, class, education, and changes in voting intention

Since the 1960s there has been a striking shift in traditional party allegiances in the US, the UK, and Australia, in which blue-collar workers, many of them socially conservative, and anxious about immigration, race, sexuality, and gender issues, began voting for Republicans, Conservatives, or Liberals, while many professionals, doctors, lawyers, and academics with high incomes increasingly voted for the Democrats, Labour (in Britain), or Labor (in Australia).

Andrew Robb, then federal director of the Liberal Party, later a minister and lobbyist, told the Sydney Institute that in the 1996 election the Coalition had won 47.5 per cent of blue-collar workers' votes (to Labor's 39 per cent) and 47.0 per cent of the Catholic vote (to Labor's 37 per cent), and the majority of votes of women, 'Anglos', the aged, and the young.

As society changes, so do voting habits.

Changes in the voting pattern of Catholics are particularly significant.

Until 2020, the United States had elected only one Catholic as president, John F. Kennedy, in 1960, and one vice president, Joe Biden, in 2008 and 2012. However, every president in recent decades has had—or claimed to have had—a strong religious faith.

In the United Kingdom, there has never been a Catholic prime minister, although Tony Blair became a convert in 2007, after leaving office. The subject of personal religious beliefs has never been an issue in elections.

By contrast, Australia has had eight Catholic prime ministers. James Scullin (ALP) was the first, in 1929, immediately followed by Joseph Lyons in 1932.

When Scullin's government split over ways of handling the Depression, the original United Australia Party (UAP) was formed. Lyons defected from the ALP in 1931 to head the new party, and was the first Catholic to lead the conservatives.

Australian prime ministers since 1929 and their religious adherence	
James Scullin (1929–32)	practising Catholic
Joseph Lyons (1932–39)	practising Catholic
Earle Page (1939)	nominal Anglican
Robert Menzies (1939–41 1949–66)	nominal Presbyterian (brought up Methodist)
Arthur Fadden (1941)	nominal Presbyterian
John Curtin (1941–45)	lapsed Catholic
Frank Forde (1945)	practising Catholic
Ben Chifley (1945–49)	nominal Catholic
Harold Holt (1966–67)	agnostic
John McEwen (1967–68)	nominal Presbyterian
John Gorton (1968–71)	agnostic
William McMahon (1971–72)	practising Anglican
Gough Whitlam (1972–75)	agnostic
Malcolm Fraser (1975–83)	nominal Presbyterian
Bob Hawke (1983–91):	agnostic
Paul Keating (1991–96)	nominal Catholic
John Howard (1996–2007)	practising Anglican (brought up Methodist)
Kevin Rudd (2007–10; 2013)	practising Anglican (brought up Catholic)
Julia Gillard (2010–13)	atheist
Tony Abbott (2013–15)	fervent Catholic
Malcolm Turnbull (2015–18)	Catholic convert
Scott Morrison (2018–)	fervent Pentecostal

After Lyons died in 1939, a succession of UAP-Liberal Party leaders followed who were either nominal Christians or agnostic, except for John Howard and Alexander Downer, both practising Anglicans.

Religion used to be a deeply divisive issue in Australian

politics, with conservative parties drawing support from Anglicans, Presbyterians, and most mainstream Protestant churches, while Labor—until the party split of 1954–55—overwhelmingly won the votes of Catholics. The conscription referenda during World War I and deep controversies about the British role in suppressing Home Rule for Ireland led to bitter sectarian divisions, both social and political.

During Menzies' second term as prime minister, cabinet was dominated by Presbyterians and Freemasons, but there was always one token Catholic minister—just as there was one (also token) woman minister.

After the Labor split over attitudes to communism, many Catholics broke away to support the Democratic Labor Party (DLP), which faithfully provided second preferences that secured the Coalition's record 23 years in office (1949–72). Menzies, who saw the breaking down of the sectarian divide as one of his major achievements, set the precedent of providing state aid to Catholic schools, and this became a very important factor in cementing the Catholic–conservative link.

Bob Santamaria, who was the inspiration for the Labor split, a skilled communicator, and often a strategic thinker, made a fundamental error. By sequestrating Catholics within the DLP, he effectively denied them the opportunity to seek election in one of the mainstream parties. (The DLP elected some senators, under proportional representation, but never won a seat in the House of Representatives.)

The DLP ended, in effect, after Whitlam and Labor won in 1972, defeating the Coalition + DLP bloc. Catholics were then given a leave pass, which they took advantage of, moving into the Liberal Party, the National Party, and the ALP, where they made up a significant core of the Right faction.

The first Catholic to be elected leader of the National Party (formerly the Country Party) was Tim Fischer, in 1990. Since then, the National Party has had three Catholic leaders: Mark Vaile, Barnaby Joyce, and Michael McCormack. They were not elected *because* they were Catholics: the significance of the change is that once they would not have been possible contenders.

In John Howard's first government (1996–98), the school with the largest representation in cabinet was Xavier College, Melbourne.

Labor leaders Bill Shorten, a lifelong Catholic, became an Anglican, and Bill Hayden, a lifelong agnostic, converted to the Catholic 'faith of his fathers' (strictly, in his case, his mother's).

Since Menzies' introduction of state aid for church schools, originally at a modest scale to fund science classrooms, it has expanded greatly, with bipartisan support. Church and/or independent schools have become multimillion-dollar international businesses, with strong links in south-east Asia. The rivers of gold provided by taxpayers for non-state schools—and tax relief for parents—are important factors in maintaining political support for the Coalition, and many voters believe (correctly, I think) that the Coalition is committed to further increases and that Labor is very uneasy about it.

Alongside these political and financial realities, two paradoxes have arisen:

- Despite Australia becoming increasingly secular, there has been a significant rise in the number of political leaders with strong religious convictions; and
- As churches and church schools come under increasing scrutiny and condemnation for malpractice, their tax-exempt status as charities and their privileged postion as recipients of taxpayer-funded support continues.

Disengagement

In 1950, when Australia's population was 8.2 million, Robert Menzies' Liberal Party had 197,000 members. In the December 1949 federal election, which Menzies won handsomely, 4.9 million citizens were enrolled, and the total Liberal primary vote was 1,813,748. So, Liberal branch members equalled 4 per cent of all enrolments and just on 11 per cent of the aggregate Liberal vote.

By 2020, party membership as a proportion of overall voting enrolments had fallen dramatically to a tiny fraction of what it had been 70 years earlier. Today, political parties are rather shifty about revealing their membership numbers. This is due, in part, to their embarrassment about very low levels of membership, but also because, as a result of branch stacking and the recruitment on a heroic scale of people representing a sectional interest, it becomes hard to distinguish between the quick and the dead, double counting, and over-optimistic projections. Even party officials may not know the correct figures themselves while making brave assertions about them.

As the purpose of a general election is not just to choose a local MP but to elect a government, the great majority of voters, understandably, see no alternative to the major parties. However, on polling day many perform their civic duty with pegs on their noses—and they have no interest in joining parties. (In May 2019, 5.5 per cent turned up but voted informally.)

Our major parties claim to have a total membership (on paper anyway) of about 100,000—that is, about 0.66 per cent of voters. In reality, it is more likely to be less than 30,000 (0.2 per cent), not all of whom will know that they hold party tickets. Many will have been signed up *en bloc* by some factional interest, with the numbers then used to negotiate in trading for patronage and preferment. Perhaps 15,000 in total could be regarded as activists.

By contrast, the total membership of sporting, especially

football, clubs would be well above 1 million.

In 2017, when he was president of the Victorian Liberals, Michael Kroger complained that the total membership of the state branch was only 13,000, with half of them aged above 70. In New South Wales, Liberal Party membership was even lower—possibly less than 10,000.

In 2014, two figures were provided for the membership of the ALP: 53,930 by party officials, and 44,000 by Bill Shorten, then leader. Both figures are likely to have been inflated. (In the postal ballot by branch members in May 2018 for the election of the ALP national president, the result was published only in percentages, and not total figures.)

In *Crikey*, on 18 July 2013, Cathy Alexander wrote, 'There are more people on the waiting list to join the Melbourne Cricket Club than there are rank-and-file members in all Australian political parties put together.' The MCC figure was 232,000. (In 2020, this proposition would still have been correct.)

She went on, '*Crikey* found that Freemasons were far more open and transparent on their membership [figures] than the Liberals, Nationals or Labor.'

In practice, an active and strategically placed minority can exercise far more power than a large but uncoordinated majority.

None of the major parties shows any enthusiasm for recruitment from the broader community on a large scale, particularly if it were to threaten the power base of existing factions. So, for example, there has been selective recruiting of Mormons and evangelicals in the Victorian Liberal Party, to shore up preselections for conservative candidates. As a result, 1,000 Mormons and evangelicals can now determine the outcome of contests in three or four federal electorates.

Within the Coalition parties, differences are often highly personal—and extremely bitter, although baffling to outsiders.

Little information has been published about Liberal Party

stacking, but there are some striking examples. When Malcolm Turnbull defeated the sitting MP Peter King for preselection in the very safe New South Wales electorate of Wentworth in 2004, both sides stacked on a heroic scale, adding more than 3,000 new members in a few weeks. Turnbull boasts about this in his memoir *A Bigger Picture* (2020).

Scott Morrison's preselection for the electorate of Cook in 2007 was distinctly odd. He lost heavily in the first ballot, then the winner was forced out after a series of anonymous attacks, and Morrison won after a second ballot was called.

Labor Party factions recruit in a different way. For example, imagine that a Ruritanian Soccer Club with 300 members in operating in one of Melbourne's western suburbs. A suggestion is made that if the Ruritanians sign up as ALP branch members (thereby bolstering the numbers for a particular faction), some generous donor will pay for their subscriptions. All the 300 need to do is turn up to vote once each year, often arriving together by bus, where they will be told what to do to elect conference delegates, Oh, and by the way, the local (ALP) council might be interested in improving the club's facilities, and—even better—perhaps a job could be found in an MP's office for the club secretary's daughter. Sound attractive? Indeed. And not even illegal.

Adem Somyürek, a Victorian ALP minister until he was sacked in June 2020, was plucked from obscurity overnight by revelations in *The Age* and on Channel 9's *60 Minutes* about 'stackathons' that he had organised for his 'moderate' sub-faction of the Victorian Right—which had some surprising personal linkages to influential groups in the Left.

In practice, the terms of 'left' and 'right' have become meaningless and cut off from their historic origins. They are now used for badging, as with the familiar names of football clubs such as 'Magpies' or 'Demons'.

Somyürek claimed that his 'moderate' sub-branch of the Victorian Right had been 'stacking' (a.k.a. 'energetic ethnic recruiting') 'on an industrial scale', and that two-thirds of the party members in the state had been corralled (and their subscriptions paid) by him and his backers.

If he was stacking 'on an industrial scale', it was essentially inside a cottage industry, because the total numbers were so small. The Victorian ALP claimed to have 16,000 members. This may well have been the number of names on the books, not all of whom would necessarily have been alive or even aware of their membership. The number still breathing, active, and involved would have been far smaller—perhaps only one-third.

Somyürek's claim that two-thirds of the 16,000 (that is, 10,666) were under his influence was challenged by party officials, who accused him of gross exaggeration. The correct figure, they said, was only 4,000 (25 per cent).

Nevertheless, the Somyürek affair raises three serious issues.

Many senior Labor figures have, or claim to have, serious memory deficits, protesting that evidence of branch stacking is new and distasteful to them—indeed, shocking. The ALP National Committee of Review, conducted by former Labor leaders Bob Hawke and Neville Wran in 2002, accepted factions as a fact of life, but condemned 'branch stacking, and the cancerous effect this activity has on the democratic traditions that have been the strength of our Party'. The ALP National Review Report by former state Labor leaders Steve Bracks and Bob Carr, and former New South Wales senator John Faulkner in 2011 also examined the decline in 'volunteerism' and the resort to branch stacking, but this part of their report was not published. So it is not a new phenomenon.

Surprisingly, notwithstanding its murky internal processes, the Victorian ALP has had a series of very able and principled leaders: John Cain, Joan Kirner, Jim Kennan, John Brumby, Steve

Bracks, and Daniel Andrews. All but one (the late Jim Kennan) were premiers, their governments achieved major reforms, and they personally set high ethical standards. Similarly, the ALP has consistently won a majority of Victorian seats at every federal election since 1998. This suggests that voters have a high level of cynicism and detachment about both major parties ('They all do it'), which reinforces their determination not to be directly involved, even though they vote for them.

As the former Labor senator and minister in the Hawke–Keating governments John Button wrote in his *Quarterly Essay*, 'Beyond Belief' in 2002:

> The factional system values mediocrity above ability, and loyalty (to the faction) above life experience. It makes for an inward-looking rather than an outward-looking party. There is plenty of evidence that the public recognises this, but the factional leaders don't want to know about it.

The ALP once had a highly unionised support base of blue-collar workers, overwhelmingly male. Many of them worked in large factories, where solidarity and collective action were axiomatic. Now this has gone or is going, because anufacturing is in sharp decline as an employment sector. Union membership accounts for about 18 per cent of public employment (mainly teachers, public servants, and police) and 12 per cent in the private sector.

Trade union members now total only 1.5 million, of which 633,200 are classified as managers/professionals; 480,000, blue collar; and 455,100, sales/service/clerical. In the construction and mining industries, many workers, certainly the most highly paid, have become independent contractors. Trade union members in mining comprise only 18 per cent of the workforce; in construction, 10 per cent.

The unions are locked into the ALP as part of the factional system, and some leaders have become, in effect, patronage traders—witness, for example, the scandals in the Health Services Union associated with union officials Kathy Jackson and Craig Thomson.

This is not typical, of course. The leadership of the very large 'Shoppies' union (The Shop, Distributive and Allied Employees Association, or SDA) became obsessed with ethical, but not industrial, issues (opposing same-sex marriage, stem-cell research, and Aboriginal land rights). This emphasis is changing, but the SDA has maintained its clout in Labor factions and the caucus.

Paradoxically, as membership of trade unions *declines* in the work force, trade union control of the ALP through the factional system *increases*, and, in a series of stability pacts, rank-and-file branch members are effectively shut out of policy-making and the selection of candidates.

When I began attending ALP conferences more than half a century ago, most of the delegates (virtually all male, I must admit) were actual workers. Many had suffered from industrial injuries, and it was common to see delegates who were missing an eye or fingers, or had suffered burns. The great majority had left school early, but had several elements in common: they were passionate readers, had strong convictions, and knew how to debate a powerful case. Some had been jailed for their political/industrial activity.

In state and federal conferences these days, 50 per cent of delegates are elected by (heavily stacked) branches, and the other half represent affiliated trade unions. They are chosen by head office—not directly elected by their members, such as shop assistants or timber workers, few of whom would belong to the party.

In fact, at recent ALP conferences, one would be more likely to find a platypus than a blue-collar worker. The overwhelming majority of delegates owe their livelihood to the ALP or its trade union affiliates. They take instructions from their factional leaders.

Many, probably most, are tertiary educated, but have no experience in arguing a case: they just 'stay on message'. Typically, after voting concludes at ALP state conferences, delegates go home. The idea of being imprisoned for a cause would give them the vapours.

The career is the driving force for budding apparatchiks: first in a union, lobbying organisation, or MP's office, then in parliament, and then, over the horizon, in consulting or lobbying. This is true of both the hegemonic parties.

As a professorial fellow at the University of Melbourne, I regularly receive requests for advice from young people who tell me, 'I'm thinking of going into politics.'

I ask, 'Which party?'

Typically, they respond, 'I don't know. I haven't decided yet.'

Then I ask, 'So, you're passionate about an issue?'

Mostly, I receive a quizzical look and the response, 'Oh, should I be?'

Clearly, many young people see politics as an alternative career path to accountancy, or teaching, with no necessary commitment to ideas or debate.

Usually, members of parliament are drawn from a very narrow gene pool, and follow a depressingly similar career path: they begin in student politics; then, after graduating, they take a party/union/corporate/lobby-group organising job; then they become an MP or senator; then they become a minister; then they 'resign to have time with the family'; and then they head for the golden land as a lobbyist/board member/employee (in areas such as gambling, banking, Chinese interests, and mining). Not surprisingly, the hegemonic parties have shown a marked incapacity to tackle major, complex, intractable ('wicked') problems and matters of conscience.

In the House of Representatives elections, the Greens have been disadvantaged so far, because their vote is spread Australia-wide. The Nationals, fighting on their own turf, benefit from a geographically concentrated vote. For example, in the May 2019 election, the Greens outpolled the National Party by a considerable margin in the aggregate, winning a total of 10.4 per cent nationally, but securing only a single seat (Adam Bandt in Melbourne). With an aggregate vote of 6.4 per cent, only 60 per cent of the Greens', the Nationals won sixteen more seats in the House of Representatives.

However, in elections for the House of Representatives, despite the manifest failure of other parties on environmental issues, the Greens' national aggregate of first-preference votes has plateaued: 11.8 per cent in 2010; 8.7 per cent in 2013; 10.2 per cent in 2016; and 10.4 per cent in 2019.

It might have been expected that the Greens would poll far better in the Senate, where voters can give themselves the luxury of casting an 'expressive vote', not to form a government, but to indicate their concerns about the environment and about growth being seen as an end in itself. But there is little difference in the results the Greens gain there: 12.9 per cent in 2010; 9.2 per cent in 2013; 8.7 per cent in 2016; and 10.2 per cent in 2019. Oddly, there has never been more than a single Greens senator in New South Wales.

Leadership and 'authenticity'

The veteran British Labour politician Tony Benn argued, persuasively, that leaders generally fall into three categories: the mad man (or woman), who breaks the rules and attempts what seems impossible; the straight man, who is transparent, persistent, predictable, and well organised; and the fixer, who is versatile, opportunistic, and mercurial.

I assume that he was borrowing from Homer. There are three outstanding models of masculinity and leadership in *The Iliad* and *The Odyssey*: Achilles, Hector, and Odysseus.

Achilles, the greatest Greek warrior, is the model for the ruthless individual who will let nothing stand in the way of his vision—testosterone-fuelled, trampling on rivals, determined to prevail. He refuses to accommodate another point of view, and exemplifies the mad-man stereotype.

Hector, son of Priam, king of Troy, is a devoted husband, father, and son; a reluctant warrior, conservative in his dedication to order, tradition, and institutions; brave, but with a conscience, uneasy about using force; and determined to improve his culture, flexible, and open to new ideas. He is the archetypal 'straight man'.

Odysseus (Ulysses in Latin), a Greek, king of Ithaca, is adaptable, a ferocious warrior where necessary, but essentially versatile, devious, a liar, imaginative, a great storyteller, with an elastic morality, ideally suited (in the contemporary world) to be a diplomat, property developer, or politician. He epitomises the classic 'fixer'.

Using Benn's categories, this is how I would list a clutch of recent leaders in the US, the UK, and Australia:

- *Mad Man (or Woman)*: Winston Churchill, Margaret Thatcher, Gough Whitlam, Paul Keating, Mark Latham, Kevin Rudd, Tony Abbott, Donald Trump, Boris Johnson.
- *Straight Man*: Ben Chifley, Robert Menzies, Clement Attlee, Gordon Brown, Bill Hayden, John Howard, Malcolm Fraser, Julia Gillard, Anthony Albanese.
- *Fixer*: Bob Hawke, John Howard, Tony Blair, Bill Clinton, Bill Shorten.

I happily include myself in the 'mad' category, because I was always aiming for objectives that were seen as beyond the reach of conventional politics, although I was strikingly deficient in the killer instinct.

The classification probably needs some revision. John Howard, for example, was a mixture of Straight Man and Fixer.

Malcolm Turnbull, as mentioned earlier, is very hard to classify. So is Scott Morrison.

Much was expected of Turnbull when he took the prime ministership from Abbott in September 2015. But he had entered into a Faustian bargain with the National Party: to lead by *not* leading, and to abandon his own long-held policy commitments. If the document setting out his commitments to the Nationals had been published, it would have blown up the Coalition.

Both Turnbull and Shorten were able, energetic, widely read, and masters of detail, but oddly lacking in the capacity to inspire or persuade. Shorten was an enigma. He had many characteristics of the Fixer, but I have no clear idea what is inside. (He may not, either.) Perhaps the dominant factor for both Turnbull and Shorten was the consuming ambition to be prime minister.

Many voters appear to set a high premium on politicians showing authenticity, without necessarily endorsing their policies. Past experience shows that voters respect leaders who are prepared to tell them what they don't want to hear—and then explain why.

The mind boggles at the thought of federal and state Labor leaders Gough Whitlam, Paul Keating, Don Dunstan, or Bob Carr turning up at a football match, except at gun point. Their passions for Graeco-Roman ruins, Mahler's symphonies, chamber music, and Marcus Aurelius were not majority interests, but they were accepted as being authentic. They did not attempt to ingratiate themselves as just being part of the mob. They were not.

Plebiscitary democracy

On the face of it, what could be more open and democratic than a referendum or plebiscite? Every voter is asked directly to express her/his opinion on a contentious subject, rather than leaving it to a group of self-serving parliamentarians, remote from real-life experience, inhabiting a political bubble. (Canberra? Westminster? Washington?)

And yet, in practice, theere is an inbuilt tension between parliamentary democracy, where the mandate to govern has to be renewed every two, three, four, or five years, and a plebiscitary democracy, which makes a decision that—some would argue—should apply until the end of time.

With constitutional referenda in Australia, there is always an advantage for the 'No' vote, compounded by compulsory voting.

For the 'Yes' vote to succeed, it must receive both an absolute majority of votes and gain the support of a majority of states. (That is, four of the six.) A 'Yes' vote means essentially, *I have thought about the issue and I am convinced*, while a 'No' vote can mean two things: *I have thought about the issue and I am opposed to it*, or *I don't know. I haven't thought about the issue and I couldn't care less, but since I have to turn up, I'll vote 'No'.*

Referenda are very different from elections. In an election there is always a winner (not always the voting public). But with referenda, 'don't knows' will often determine the result, especially with compulsory voting.

In 1937, a referendum to give the Commonwealth power to make laws over aviation and air navigation was defeated because, although 54 per cent of Australians voted 'Yes', the proposal was carried in only two states (Victoria and Queensland—both by large margins). The High Court subsequently allowed the Commonwealth to legislate on aviation by using the external affairs power in the Australian Constitution (s. 51 (xxix)) to ratify international treaties.

In May 1967, 93.7 per cent of Australians voted to allow Aborigines and Torres Strait Islanders to be counted in the census (which had been a striking omission in the 1901 Constitution), and to have the option of enrolling to vote.

Referendum proposals providing for elections for the Senate

and the House of Representatives to be held on the same day—
the invariable practice since 1974, and in most elections between
1901 and 1953—have been put four times (1974, 1977, 1984, and
1988) and defeated each time. In 1988, a referendum proposal to
establish fair elections in the states failed to carry a single state.
Another referendum question in 1988 to entrench trial by jury for
indictable offences, to extend freedom of religion, and to ensure
fair compensation for people whose property was acquired by
government secured only a 30.8 per cent 'Yes' vote.

In April 1977, 80.1 per cent of Australians voted 'Yes' for a
proposition to impose retirement ages upon federal judges, and
legislation subsequently set the age at 70. As a result, High Court
justices Murray Gleeson, Michael Kirby, and Kenneth Hayne left
the court at 70, and then went on to other things, with no sign of
senescence.

In November 1999, two propositions were put to the Australian
electorate: for Australia to become a republic and for the Constitution
to be amended, inserting a preamble with a brief acknowledgement
of democracy, 'upholding freedom and tolerance, individual dignity
and the rule of law, honouring Aborigines and Torres Strait Islanders
… recognising the nation-building contribution of generations of
immigrants, mindful of our responsibility to protect our unique
natural environment'.

The republic proposal was rejected by 55 per cent to 45 per cent,
and the proposed preamble was defeated far more decisively, by 61
per cent to 39 per cent.

There was a curious alliance between monarchists and 'direct
election' republicans in the 'No' campaign. Both opposed the
bipartisan model, which provided for the governor-general to be
replaced by a president who would act as an apolitical head of state,
and to be chosen by a joint sitting of the parliament, with a two-
thirds majority being required. This would inevitably have resulted

in the choice of a consensus figure and would have excluded a partisan.

The monarchists thought the proposition was too republican. The 'direct electionists' thought it was not republican enough, insisting that if the 'Yes' proposition was defeated, it would be quickly replaced by a new republican model.

More than twenty years on, there is no sign of this happening.

The result cut across party lines completely. The prime minister, a vehement opponent of a republic, was presumably mortified when his electorate of Bennelong (in New South Wales) recorded a 54.6 per cent 'Yes' vote. The Labor opposition leader, Kim Beazley, a supporter, was taken aback when his electorate of Brand (in Western Australia) voted 'No' by 66.3 per cent.

My estimate is that of the 55 per cent of Australians who voted 'No', perhaps 30 per cent supported a monarchy, 10 per cent were direct-election republicans, and 15 per cent were simply not interested, and would not have voted in a non-compulsory poll.

Public-opinion polling indicates a strong correlation between income and voting in referenda—with people on high incomes voting 'Yes', and those on low incomes voting 'No'. There is also a strong correlation between levels of education and age in a referendum.

New Zealand provides an admirable model for conducting a referendum on a complex issue, as a result of having changed from a 'first past the post' voting system to a 'mixed-member proportional' (MMP) system, as used in Germany. (This had been recommended in 1986 by a royal commission set up by Labour prime minister David Lange's government.)

Under the MMP model, every elector has two votes: one for an MP to represent a locality; the second, for a 'list' (or 'lust', as the New Zealanders engagingly pronounce it) that represents a national spread for the parties. The MMP system was adopted in principle in 1992, was confirmed by a second referendum in 1993, and once

more in 2011. The highly appealing feature of MMP is that—subject
to reasonable thresholds being met—if a party achieves a certain
percentage of the vote, it will win precisely that percentage of seats.

The best feature of the New Zealand referendum system is that
every household receives a briefing document prepared by the clerk
of the parliament which sets out in balanced fashion—with excellent
diagrams—the case for 'Yes' and 'No', and provides answers to the
questions that are likely to be asked about costs and implementation.
No hysteria. No misrepresentation. Just a balanced view.

Sometimes a plebiscite can work out very well—for example, in
2017, with Australia's voluntary postal survey on same-sex marriage
(and a similar vote in Ireland). This was an example of the political
class having previously failed to act because it feared that the general
community was not prepared for change. There was a 79.5 per cent
response, and 61.1 per cent of respondents voted 'Yes'. Then the
parliament got on with legislating the details.

Again, the voting returns cut across party lines. Of Australia's
151 electorates in the House of Representatives, only seventeen
voted 'No'—eleven of them ALP seats classified as safe, but with
significant Muslim or African minorities.

The electorates held by Tony Abbott, Peter Dutton, Scott
Morrison, Barnaby Joyce, and Michael McCormack voted 'Yes'.

Putting one simple question in a referendum has been a
technique used by dictators and authoritarians (such as Napoléon I
and III, Hitler, Mussolini, Marcos, Erdoğan, and Putin) to augment
their power.

In July 2020, Russians voted 78 per cent 'Yes' to a single
question: 'Do you approve amendments to the Constitution of the
Russian Federation?' There were fourteen unnamed propositions,
only one of which was significant, enabling Vladimir Putin to seek
two more six-year terms after his current mandate ends in 2024.

And a simple 'Yes'/'No' option is not particularly helpful in resolving very complex issues—for example, in the case of voluntary assisted dying.

John Maynard Keynes is credited with saying, 'When my information changes, I alter my conclusions. What do you do, sir?' Parliaments are far from perfect, but they can be flexible and even principled when new evidence appears.

'Brexit' and the triumph of Boris Johnson

The European Union referendum held in the United Kingdom in June 2016 was a striking example of a complex issue being determined by a single vote on a single question. The 2016 vote resulted in a majority for Britain's exit ('Brexit', in the portmanteau word that followed) from the European Union.

In 1975, the vote for the United Kingdom European Communities Membership referendum had a majority of 67 per cent for Britain's entry. Why did that result not last forever—if the logic of the Brexiteers is followed?

The 2016 EU referendum had been instituted by prime minister David Cameron in an attempt to stare down a strongly anti-European sentiment in his own party, the Conservatives, and he was confident of a majority vote in favour of remaining. This proved to be wishful thinking.

As the philosopher A.C. Grayling reminds us in *Democracy and its Crisis*, 'all MPs and members of the House of Lords were told [in Briefing Paper 07212 on 3 June 2015] that the referendum [on Brexit] was advisory only, and would not be binding on Parliament or government ... The outcome was that 37 per cent of the restricted electorate given the franchise for the referendum voted to leave the EU.'

'It says in section 5 that the referendum is non-binding, advisory, consultative; and section 6 points out that if there were to be any suggestion otherwise, there would need to be a supermajority requirement.'[1]

The campaign was badly argued on both sides—feebly by Cameron, mendaciously by Nigel Farage and his UKIP party—and the position of Labour under Jeremy Corbyn was opaque. Within the Conservative ranks, the strongest 'Leave' campaigners were Boris Johnson, Michael Gove, and Jacob Rees-Mogg. Theresa May was a low-key 'Remainer'. The strongest feelings against the EU were from the areas with the least exposure to Europe; the most enthusiastic Remainers had the most experience of Europe.

The best-qualified potential voters (that is, Britons working in Europe) were specifically excluded. Eric Idle, a Monty Python veteran, living in the United States, was barred from voting, and commented bitterly that the Russians had more influence on the Brexit result than he did. The turnout was 72.2 per cent.

The final vote—52 per cent 'Leave', and 48 per cent 'Remain'—looks comparatively close, but it conceals deep regional/geographical divisions. The results in 'Brexitland' and 'Remainland' were dramatically different. Among very high 'Remain' votes were Edinburgh and Cambridge (74 per cent each); Oxford (70 per cent); Scotland generally (62 per cent); London (60 per cent); Manchester (60 per cent); and Northern Ireland (56 per cent). The Midlands, Yorkshire, the East, and the North voted around 56 per cent for Brexit, with some constituencies above 70 per cent.

Statista Research reported that voters aged 18–24 supported 'Remain' by 73 per cent; voters aged 55 and over supported 'Leave' by 59 per cent; those with a degree voted 74 per cent to remain; and those with no tertiary qualifications voted 65 per cent to leave.

The Russians were eager to see Britain leave the EU, and hacking was useful in targeting susceptible voters.

After the vote, David Cameron disappeared from political office, and the caveat about the vote having been 'non-binding, advisory, consultative' mysteriously went with him. Then followed a three-year clash between 'plebiscitary democracy' ('No' means 'No') and attempting to deal with the complexity of unexplored issues (Ireland being just one) in the 'parliamentary democracy' model.

Current Brexiteers are hypocritical and inconsistent. They regard the 2016 'advisory' vote as absolutely binding for all time, and not to be challenged or tested by a second vote, but dismiss the 1975 referendum, which their parents voted in, as irrelevant and non-binding.

Theresa May succeeded Cameron as prime minister; was humiliated by Donald Trump, who asked her to appoint Nigel Farage as ambassador to Washington; and failed to negotiate an exit package that was acceptable to the EU, the House of Commons, and Ireland over the incendiary issue of a 'hard border', which nobody seems to have thought about during the Referendum campaign. Chaos, or near chaos, reigned.

The Conservative Party membership now elects its leader from a short list chosen by the parliamentary party. After May resigned, there was a ballot of party members in July 2019. Boris Johnson was elected leader with 92,153 votes (66 per cent) to Jeremy Hunt's 46,656 votes (34 per cent), a total of 138,809. Johnson became prime minister, but every parliamentary strategy/tactic failed. Despite this, he called a general election, and in December 2019, the Conservatives won a majority of 80 with an aggregate vote of 43.6 per cent. Labour's vote fell to 32.1 per cent; the Liberal Democrats had 11.6 per cent; the Scottish National Party, 3.9 per cent; and the Greens, 2.7 per cent. The turnout was 67.3 per cent—lower than for the 2016 referendum.

How united was the kingdom? The Conservatives won England; the Scottish National Party, Scotland; Labour, Wales; and the Democratic Unionist Party, Northern Ireland.

Ultimately, Britain left the EU on 31 January 2020.

Johnson faces a paradox. His espousal of the 'Leave' cause appears to have been entirely opportunistic, and the suggestion that he tossed a coin to decide which side to support on the EU issue may even be true. He now faces a challenge about how to unite the UK. In the Brexit debate, he was saying to his followers, *We are better off alone.* Now he wants to tell Scotland, *You won't be better off on your own.*

Perhaps he can get away with it. Conservatives are very forgiving about policy failures in their party.

The Conservatives and Labour

The British Conservative Party has been one of the great success stories in modern politics. Since its foundation in 1834, its leaders have held the prime ministership for 94 years. And yet it was the Conservatives' good fortune that its great historic campaigns all failed: for the Corn Laws, for child labour, and for the House of Lords' power to veto legislation; and against Catholic and Jewish emancipation; the secret ballot; manhood suffrage; votes for women; Home Rule for Ireland; and independence for India and Pakistan.

It is hard to imagine modern Britain if all these policies had survived.

The Conservative Party has succeeded by *not* proposing bold ideas, and by *following* public opinion, rather than by leading it. Nevertheless, in the twentieth century, under universal suffrage for most of it, the Tories held office far longer than Labour and the Liberals combined.

The only prime minister of Jewish heritage was Benjamin Disraeli, a Conservative. Britain's only women prime ministers have been Margaret Thatcher and Theresa May, both Tories. Oddly, despite the white racist elements in the Conservatives, it has 22 MPs, and some peers, of Asian or African descent, including two successive chancellors (Sajid Javid and Rishi Sunuk).

The Conservative Party now has about 160,000 members. The party membership is 71 per cent male and 38 per cent aged over 66. Sixty per cent want to bring back hanging—a far more reactionary view than that of Tory voters across the community. In the 1950s the Conservatives reached a peak of 2.8 million members.

The British Labour Party has a membership of 580,000—the largest of any party in Europe, many of them zealots, recruited when Jeremy Corbyn was leader. Again, the relatively large membership is not necessarily an indicator of public support; in fact, in the general election of December 2019, Labour secured its lowest popular vote since 1935.

The Royal Society for the Protection of Birds has 1.1 million members—more than all Britain's major political parties combined.

Toxicity

In recent decades, political life in Canberra has become toxic, with a breakdown in personal relationships across party lines, and a recourse to incessant personal attacks, wild exaggeration, and the endless repeating of slogans, with politicians having abandoned the practice of debating with ideas and of using sentences that contain verbs. People with long political experience, many on the Coalition side, volunteer that the Abbott and Morrison governments have been the most vindictive they can recall, although Howard had some form there, too. Supporters are rewarded and opponents punished in unprecedented ways.

Kevin Rudd appointed Brendon Nelson and Tim Fischer, former Howard ministers, as ambassadors, extended Amanda Vanstone's term in Rome, and chose Robert French, briefly a Liberal Party member, as chief justice of the High Court, in preference to Jim Spigelman, who had worked for Whitlam. By contrast, the Morrison government's appointment of Gary Gray, a former ALP national secretary and minister, later a company director, as ambassador to Ireland was a rarity.

Paradoxically, the toxicity in parliamentary proceedings, always timed for maximum media exposure and commentary on the 'dark web', is a by-product of trivialisation. The Australian political system has proved incapable of tackling major problems, such as climate change, the refugee issue, the ethical basis of taxation, huge disparities in life expectancy, education, the delivery of health services, and secrecy and corruption in public life. Ministers and members shout at each other, and look for cheap applause, because they have no wish to debate public issues and have only two preoccupations: personal career advancement and winning the next election.

Racism and spin

Race and inequality, like the unconscionable treatment of asylum seekers, are uncomfortable issues that most Australians would rather not think about.

On 25 May 2020, George Floyd, an African-American, was the victim of an extra-judicial execution by a white police officer in Minneapolis, Minnesota, captured on social media for an excruciating eight minutes and 46 seconds, and viewed globally. Floyd had been picked up—and knocked down—by police, on suspicion of having tendered a counterfeit $20 bill.

As he lay dying, with the policeman's knee on his neck, Floyd repeated, 'I can't breathe.'

These words were used on banners in 'Black Lives Matter' demonstrations throughout the world, at a time when coronavirus pandemic restrictions on public gatherings were being enforced. Demonstrators had to choose between making a public demonstration of solidarity or risking infection and/or fines or arrest by police.

The killing of African-Americans by police has been completely disproportionate to their numbers in the community, and so frequent as to become almost routine. But seeing Floyd's slow-motion death, and the casual reaction of other police watching it happen, provoked an international reaction, the strongest for decades.

In many American cities, heavily armed police have been playing a paramilitary role—and in some cases provided with tanks. They are essentially warriors, not guardians. They shoot first, and ask questions later. The use of lasers, tear gas, pepper spray, and rubber bullets are less lethal, but more common. Their actions suggest that the American Civil War is far from over.

In Australia, Floyd's death was a catalyst for protests about race. Since the Royal Commission into Aboriginal Deaths in Custody reported—adversely—in 1991, there have been 434 more fatalities, and not one conviction of the perpetrators. Our deaths occur in private and are rarely filmed.

Scott Morrison, predictably, asserted, 'This is not the time' for protests, since COVID-19 social distancing was still in force and—equally predictably—he assured 2GB's talkback listeners that there had never been any slavery in Australia. He wriggled out of this false statement a day later, presumably after being reminded of the historical examples of 'blackbirding', the nineteenth-century kidnapping of Pacific Islanders to work in the Queensland cane fields, and the widespread use of of Aborigines as indentured labour, often in chains.

On 24 May 2020, Rio Tinto blew up rock shelters in the Juukan Gorge in the Pilbara region of Western Australia to expand an existing iron ore mine. The rock shelters, containing evidence of continuous human occupation dating back 46,000 years, were considered to be of outstanding importance by archaeologists and anthropologists.

By sheer coincidence, news of the destruction circulated while the world was reacting to Floyd's killing. 'Black Lives Matter' and 'Black Heritage Matters' had a common theme.

The destruction of the rock shelters was a disturbing demonstration of muddle, ignorance, greed, and incompetence. Implausible though it sounds, the relevant federal and state ministers were both Indigenous, related—as uncle and nephew—from different political parties, ill-informed, and ineffectual.

Although it had sponsored a film on the heritage value of the rock shelters, Rio Tinto claimed not to have fully understood their significance, and asserted that the traditional owners, the Puutu Kunti Kurrama and Pinikura people, had acquiesced in their destruction. The corporation then came up with the feeblest justification I have ever heard—that once the explosives were in place, the destruction had to go ahead. There was a later revelation that the company knew what it was doing all along.

The Juukan rock shelters case illustrates how badly major issues can be mishandled.

A new form of apology, developed by spin doctors, has become almost an art form. A corporation acts unconscionably, or a minister says or does something damaging. They come under pressure, and respond with what sounds like an apology but is really an evasion, just a form of weasel words: 'We/I very much regret if our/my actions have caused distress.' There is no apology for the damage caused (which might lead to legal action), no accountability, and the use of the conditional 'if', purporting to showing a fine sensitivity about hurt feelings, often accompanied by a sub-text: *Toughen up!*

Decline in trust

A clusterfuck, to use the technical term, of trust-destroying factors has damaged the confidence felt in government and major institutions.

Churches, like political parties, are losing numbers, commitment, and moral authority, and have been shaken by revelations—especially in the Royal Commision into Institutional Responses to Child Sexual Abuse—of the acceptance of the sexual abuse of children. For decades, the churches' reactions have often been to protect the institution and disregard the victims.

Some political leaders act as if all values have a dollar equivalent, that forests are essentially woodchips on stumps, and that the value of a tree is as lumber, disregarding aesthetic factors or its contribution to clean air. The current obsession is that if projects will make money for somebody—for example, by grazing in national parks, or drilling for oil, or dumping mine tailings near the Great Barrier Reef, or logging in World Heritage sites, or exporting live animals, often under unspeakable conditions—they should go ahead. The appeal of money and growth in the Gross Domestic Product are irresistible, with a refusal to contemplate the downside. In the case of duck shooting, state power is entirely behind the shooters, and against the ducks. The need for more cars on more freeways outweighs the values associated with parks. Recreational shooters and four-wheel drives are now welcomed in New South Wales national parks.

Much of the mainstream media emphasises partisan advocacy and shock-based entertainment, and reinforces prejudice rather than provide information or carry out disinterested investigative reporting. This is especially true of the Murdoch empire, News Corporation, which is best thought of as a political organisation that employs journalists.

The protection and preservation of the planet and its biodiversity, for the long term, is a deeply moral matter, which we ignore in our pursuit of greed. The relief of poverty is one thing, but consumption is not an end in itself. The decay of formal religion and the long-term decline in churchgoing intensifies the need to stimulate debate and understanding about values, including the transcendental and the numinous.

Redefining politics

We must redefine politics—and grasp its importance, not just at election times. I have made an attempt, and while my definition does not exactly roll off the tongue, it captures the essence:

> Politics is the fault line between tectonic plates in society, and the electoral struggle is an expression of, or a metaphor for, unresolved, often unspoken, divisions within society—race, class, gender, religion, region, language, education, sexuality, consumption patterns and time use, self-definition and the expression of individual differences/aspirations (both positive and negative), offering a choice between different moral universes.

Tackling complex problems demands complex solutions that cannot be reduced to parroting a few simple slogans. Most of all, we need a higher level of citizen involvement in the whole process of public debate, instead of leaving it all to the political professionals.

CHAPTER NINE

The Death of Debate: the loss of language and memory

In the decade 1966–75, Australian politicians were often well ahead of public opinion on many issues, and led important innovations and reforms. Examples of this include supporting mass migration; ending White Australia; abolishing the death penalty; reforming the divorce law; decriminalising homosexual behaviour; safeguarding access to abortion; recognising the People's Republic of China; reducing tariff protection; supporting the arts; changing attitudes to the Vietnam War and conscription; creating probably the world's best national health scheme; introducing affirmative action for women; establishing needs-based education; ending censorship; admitting large numbers of refugees; and expanding tertiary education.

However, the age of optimistic, courageous reforms is now past. In recent decades, Australian politicians have been well behind public opinion on issues such as allowing same-sex marriage, taking effective action on climate change, transitioning to a post-carbon economy, protecting the Great Barrier Reef and other heritage sites, supporting voluntary assisted dying, ending live animal exports, taking a rational and compassionate approach to refugees, and moving to establish a republic. They are fearful of antagonising powerful minorities and being 'wedged'.

Paradoxically, there appears to be an inverse relationship between the number of graduates in parliament and the quality of political debate, and it is now impossible to get a straight answer to a question, whether asked in parliament or in the media.

31st Parliament 1977–80	45th Parliament 2016–19*
127 MPs	151 MPs
Number of women:	*Number of women:*
0 (0 per cent)	43 (28 per cent)
Number with tertiary qualifications:	
ALP 16/36 MPs (44.4 per cent)	ALP 64/69 MPs (92.8 per cent)
Liberals 47/68 MPs (69.1 per cent)	Liberals 50/60 MPs (83.3 per cent)
NCP 4/23 MPs (17.4 per cent)	NPA 9/16 MPs (56.3 per cent)
	Greens 1 MP (100.0 per cent)
	Independent 1 MP (100.0 per cent)

*Statistics for the current—46th—Parliament are not yet available.

When I was first elected to the House of Representatives in 1977, there were outstanding debates, often engaged in by members without much formal education, but who shared three things: they had had life experience outside the political parties and parliament, much of it very tough; they were prodigious readers; and they understood a counter-argument and how to rebut it.

However, this was not a golden age for representing a cross-section of Australia. Federal MPs had four things in common: they were all white, all male, all well above the median age, and almost all of English, Scottish, or Irish descent.

But the times were dominated by 'conviction' politics, before the advent of 'retail' politics. I can identify about 40 MPs and senators in the 31st Parliament who were excellent debaters with a serious contribution to make.[1]

Clyde Cameron, for example, left school at fourteen, became a gun shearer, and then had a white-collar career as a union official, a member of parliament, and an unhappy minister under Gough Whitlam. But he was exceptionally well read and a very powerful speaker, with a wide range of interests, and wrote four impressive books.

When members knew that Clyde Cameron was going to speak, they would fill the chamber. In the current parliament, official notification that Deputy Prime Minister Michael McCormack was going to address the House would result in a rush for the exit.

I remember a memorable clash on the House floor between Ralph Jacobi, another South Australian autodidact, and the patrician Sydney barrister John Spender, QC, about the operation of the *Corporations Act*. Spender condescendingly complimented Jacobi on his oration, but said that there were elements of the High Court's changing interpretations that a non-lawyer would be unable to grasp. Jacobi, without a note, then analysed a series of High Court decisions, proved his point, and Spender stalked out of the chamber. I doubt if this could happen now.

In 1977, only 2 per cent of the Australian population had university degrees. The current cohort of Australians has the highest level of qualifications by far in the nation's history, with more than 27 per cent (6.9 million) holding tertiary qualifications. But this cohort, on the whole, is extremely reluctant to become directly engaged in the major issues that shape Australia's future.

Why is support for democracy as 'the best political system' declining? Why have Australians lost faith in our major institutions? Why has the level of community engagement in major policy collapsed? Why are our brightest and best disengaged from major causes? Why has the quality of political discourse, in and out of parliament, fallen to historically low levels? Why have so many Australian universities opted to become trading corporations rather than shapers of our future?

In 2020, the composition of the Commonwealth parliament—and Australian society—is radically different. More than one in four is a graduate. Parliament is approaching gender balance. And the representation of ethnic diversity is far greater than it ever was.

But the quality of debate? Who are the contemporary equivalents of the names set out in the footnote on the opposite page? I can barely identify 20 who come close.

There is policy paralysis, with a significant failure of nerve by those who purport to be leaders, largely because they have little or no grasp of how to frame an argument. Many politicians and political operatives—'apparatchiks'—have not just lost the capacity to debate: they never had it. The practice of dealing with contested ideas is unfamiliar to many MPs.

Some MPs often rely on a page of dot points that they have been handed, with no understanding of or interest in a contrary point of view, and simply declaim the material they have been given, 'staying on message' and repeating mantra after mantra *ad nauseam*.

The last serious debate in parliament on the country's involvement in war was in 1991; on arts and culture, in 1995; on the Republic, in 1998; on human rights, in 2001; on foreign policy, in 2003; and on the environment and climate change, in 2009. Neither major party will debate a fresh approach to the refugee/asylum seeker issue—leaving it to independents or Greens in two Houses to initiate action, as occurred with the Medevac vote in 2018, due to the temporary lack of a government majority. On gambling, or the surveillance state, never.

At the national level, the last debates that the ALP won were on WorkChoices in 2007 and the *Tobacco Plain Packaging Act* (2011), although the survival of the National Disability Insurance Scheme

(NDIS), even in truncated form, could be classified as a partial success. The National Broadband Network (NBN), providing a service that many developing countries would envy, originally pushed by Labor, was modified by the Coalition.

In 2009, Labor did spectacularly well in handling the Global Financial Crisis, failed to capitalise on it fully, and was later punished when Abbott, in effect, won the debate. Weirdly, in the 2019 campaign, the ALP never mentioned its economic credentials and the country's retention of an AAA credit rating during the crisis.

Labor's setting of a carbon price in 2010 was courageous, but badly argued, and was rolled back in 2013.

Labor's national primary vote has steadily declined from 43.4 per cent in 2007 to 33.3 per cent in 2020. However, ALP state governments, especially in Victoria and Western Australia, have succeeded in winning debates—and elections, sometimes on sensitive subjects such as voluntary assisted dying and the setting of ambitious targets for cutting greenhouse-gas emissions.

The ALP and the Coalition parties are oligarchic, discourage large-scale membership, reject democratic forms, and are run by factions that are essentially executive-placement agencies, while the Greens are super-sensitive and secretive about party operations.

There are factional rewards for not rocking the boat. Primary loyalty is to the faction, and 'stacking' is endemic.

Dr Andrew Leigh, the Labor MP for Fenner (ACT), is an exceptionally well-qualified economist with a PhD from Harvard. He worked at the Australian National University for six years, becoming a professor of economics, and winning the Economic Society's award for Australia's best young economist. He collaborated with the late Sir Tony Atkinson on estimating long-run Australian inequality, and has published seven books since entering parliament in 2010. Despite his capacity, and because he is not a member of a faction, he has been demoted twice in the opposition and is now

Shadow Assistant Minister for Treasury. He is very loyal to the system, and I am expressing my views, not his. He won't thank me for raising this, and he is stoically resistant to making complaint.

Parliament is no longer effective in extracting information from governments. As former High Court justice Kenneth Hayne pointed out, if we want hard evidence that can be tested, we might need to have a royal commission. We may get evidence from the courts or occasionally from a Senate committee. But from government? Not a chance.

Sitting days

Few Australians recognise that its House of Representatives holds the international gold medal for the shortest sittings of any national legislature. It is not surprising that extended debate becomes impossible: it is planned that way. Here is a select list of legislative sitting days:

Japan	150 days (average)
United Kingdom	142–158 days
Canada	127 days (average)
United States (House of Representatives)	124–145 days
Germany	104 days (average)
New Zealand	93 days (average)
Australia	67 days (average)

Governments of both major parties regard parliamentary sittings as a nuisance, taking ministers away from what they regard as their core business. They are particularly irritated by Question Time, which has become a theatre of the absurd, not a genuine search for information, in which personal attacks, gaffes, or 'gotcha!' moments are scored, as if at sporting events.

Tensions, plotting, and nervous energy are used up *within* the party room. I once asked a member of a state parliament to describe what it was like inside her/his party room. The answer was, 'The shower scene in *Psycho*.'

Australia's House of Representatives sat for 113 days in 1901, 122 days in 1904, and 44 days in 2019. The year with the lowest number of sitting days—29—was 1937, but 2020 is likely to be even less.

In my years in Canberra, the highest number of sitting days was 79 in 1986; the lowest, 38 in 1990.

'Winner take all'

Governments operate on the principle of 'Winner take all'. All opposition arguments are discounted. In the current Coalition government, ministers Matt Canavan and Angus Taylor come very close to saying, *We won the election, and that means that voters have rejected the scientific evidence on climate change.*

In practice, a 51 per cent vote equals 100 per cent power. A 180-degree change in direction with profound long-term implications can be determined by chance factors, or even by accident, in a tight race (and the horseracing analogy is appropriate).

Elections in single-member constituencies may be determined by one or two people in each 100 changing their minds on or before polling day—perhaps for very sound reasons, or perhaps on a whim. We don't know, and we'll never know. What is clear is that in the 2019 election Labor polled better with the 58.2 per cent of voters who turned up on election day, less well with the 31.8 per cent who cast a 'pre-poll' vote, and worse still with the 10.0 per cent of postal voters.

Our politics is dominated by the infantilisation of debate and the fallacy of the false antithesis ('If we don't beat up refugees, we will be overrun'; 'Too much tolerance will destroy our traditional values'; 'If we don't promote the use of coal, next month's power bill will ruin us'; 'You can't preserve the rule of law in an emergency'; 'We have to impose secrecy to protect an open society'; and 'If we increase our emission-reduction targets we won't be able to watch night footy!')

Australian politics is unusually brutal, and the killing season occurs during parliamentary sessions, which probably explains why prime ministers keep them short.

No Australian prime minister except Menzies in 1966 (and Barton, up to a point in 1903, when he became a High Court justice) left the prime ministership at a time of their own choosing: the rest were victims of election defeats, coups, or death.

There also seems to be an iron law in politics: since 1913, the first opposition leader chosen after there has been a change of government *never* goes on to become prime minister. (The exception, in 1913, was Joseph Cook, and he only lasted a year.)

Framing the debate

George Lakoff, emeritus professor of cognitive science and linguistics at the University of California at Berkeley, and author of *Don't Think of an Elephant: know your values and frame the debate* (2004), has made a profound contribution to understanding how people adopt policy positions, even when they are apparently contradictory.

In 'Understanding Trump' (23 July 2016), written before the presidential election, he wrote:

How do the various policy positions of conservatives and progressives hang together? Take conservatism: What does being against abortion have to do with being for owning guns? What does owning guns have to do with denying the reality of global warming? How does being anti-government fit with wanting a stronger military? How can you be pro-life and for the death penalty? Progressives have the opposite views. How do their views hang together?

The answer came from a realization that we tend to understand the nation metaphorically in family terms: We have founding *fathers*. We send our *sons* and *daughters* to war. We have *homeland* security. The conservative and progressive worldviews dividing our country can most readily be understood in terms of moral worldviews that are encapsulated in two very different common forms of family life: The Nurturant Parent family (progressive) and the Strict Father family (conservative).

What do social issues and the politics have to do with the family? We are first governed in our families, and so we grow up understanding governing institutions in terms of the governing systems of families.

Lakoff has long argued that the outcome of policies can be determined by how they are framed.

We can see this at work in Australia. If an estate tax is described as a 'death tax', a carbon price as a 'carbon tax', accountability as 'playing the blame game' or 'red tape', environmental safeguards as 'green tape', acting responsibly as 'virtue signalling', supporting multiculturalism as 'political correctness', and keeping coal-mining going as 'technology neutral', then the framers have reset the terms of a potential debate and are on the cusp of winning it.

Words become weaponised.

John Howard was particularly skilful at framing issues, for example 'counter-terrorist' legislation (a.k.a. 'infringement of civil liberties'), so that any objection to his suspension of the rule of law, or even criticism of the case of David Hicks, an Australian imprisoned by the Americans in the Guantanamo Bay detention camp, was denounced as anti-American or un-Australian. When Howard set the agenda, and Labor acquiesced, the prospect of opening up an argument disappeared.

Framing can shape community discourse and understanding. People who try to kill themselves in immigration detention centres might receive some sympathy if their actions are called 'suicide attempts', but not if described as 'attention-seeking incidents'. Framing determines if we refer to 'refugees' or 'queue jumpers'; 'academics' or 'bleeding hearts'; 'accountability'; 'security' or 'rigidity'; 'insecurity' or 'flexibility' (in employment); 'strategic withdrawal' or 'cutting and running'; 'reform' or 'change'; and 'scientific consensus' or 'group-think'. 'Moving on' and 'closure' mean 'Don't discuss it.'

Sometimes a sensitive word such as 'rape' can be dropped into an election campaign, as Morrison did several times in 2019, in a particularly disturbing example of 'dog whistling'.

Language

In his essay 'Politics and the English Language' (1946), George Orwell argued that language must be used clearly and precisely because it is a powerful tool of political manipulation. Using slogans, and emotive but imprecise words, or chanting to arouse a mass response is, inevitably, destructive of careful examination and personal insights. The poor use of language limits our capacity for critical thought—and if most people are doing it, perhaps it is too late to change. By destroying the effective use of language, serious debate becomes impossible.

Orwell developed this powerfully in his novel *Nineteen Eighty-four* (1949), where the rulers of the totalitarian state of Oceania propose three central mantras: 'War is peace. Freedom is slavery. Ignorance is strength.'

Oceania's rulers develop an official language, 'Newspeak', intended to eliminate nuance, complexity, and a range of emotional or intellectual reactions, so that reading great authors (who wrote in 'Oldspeak') becomes pointless. Vocabulary is diminished—the word 'thought', for example, is eliminated, and complex subjects are reduced to simplistic slogans, with portmanteau constructions such as 'goodthink' and 'crimethink'. 'Doublethink' is the act of simultaneously believing two, mutually contradictory ideas, while 'doubleplusungood' replaced the Oldspeak words *terrible* and *worst*.

Orwell was prophetic here. In the 2020 US presidential election campaign, fostering ignorance was a powerful campaigning tool. Complexity could be ignored. Only one point of view needed to be considered. There was no room for doubt.

When the ABC's managing director, Michelle Guthrie, was sacked in 2018, the board's chair, Justin Milne, described it as an 'external career-development opportunity', a splendid example of Newspeak.

Umberto Eco (1932–2016), an Italian philosopher, historian, and novelist, was best known as author of *The Name of the Rose* (1980). In 1995, the *New York Review of Books* published Eco's important essay 'Ur Fascism' (that is, 'eternal' or 'primal' fascism), in which he listed fourteen general properties of fascist ideology. He argued that they could not be organised into a coherent system, because some are contradictory and typical of other types of fanaticism and authoritarianism. However, 'it is enough that one of them be present to allow fascism to coagulate around it'.

Twenty-five years later, at least five of his general properties are current, not only in the United States, the United Kingdom, and France, but (to a lesser degree) Australia:

1. Appeal to social frustration: 'One of the most typical features of the historical fascism was the appeal to a frustrated middle class, a class suffering from an economic crisis or feelings of political humiliation, and frightened by the pressure of lower social groups.'

2. The obsession with a plot: 'The followers must feel besieged. The easiest way to solve the plot is the appeal to xenophobia.'
3. The enemy is both strong and weak: 'By a continuous shifting of rhetorical focus, the enemies are at the same time too strong and too weak.'
4. Selective populism: 'There is in our future a TV or Internet populism, in which the emotional response of a selected group of citizens can be presented and accepted as the Voice of the People.'
5. Ur-Fascism speaks Newspeak: 'All the Nazi or Fascist schoolbooks made use of an impoverished vocabulary, and an elementary syntax, in order to limit the instruments for complex and critical reasoning.'

Eco was writing about fascism, but his remarks could apply just as well to the current mindless resort to clichés and repetitive slogans.

Donald Trump's use of language has been particularly significant. His vocabulary is that of an eighth-grader—and this may be central to his appeal.

Shortly before his death in 2018, the novelist Philip Roth wrote that he had never encountered 'anything like as humanly impoverished as Trump is: ignorant of government, of history, of science, of philosophy, of art, incapable of expressing or recognizing subtlety or nuance, destitute of all decency, and wielding a vocabulary of seventy-seven words that is better called Jerkish than English'.

One survey found that Trump's most frequently used words were 'winning', 'stupid', 'weak', 'loser', 'fake news', 'deep state', 'political correctness', 'the swamp', 'smart', 'tough', 'dangerous', 'bad', 'veterans', 'amazing', 'make America great again', 'tremendous', 'terrific', 'military', 'out of control', 'classy', 'sick', 'unbelieveable', 'witch hunt', and 'hoax'.

Trump had mastered the art of 'double speak', the capacity to promote contradictory ideas at the same time—for example, that having the world's highest numbers of infections and deaths from COVID-19 'is a badge of honour', and anyway the virus 'will go away suddenly, like a miracle'.

He could give and take away almost simultaneously: 'We are heading for a disaster—but don't worry about it.' One day, he said that under the Constitution he had absolute power to take action on COVID-19; the next day, he insisted that all power was in the hands of state governors.

Trump largely communicates in memes.

The word 'meme' is a useful coinage by Richard Dawkins, in his book *The Selfish Gene* (1976), defined as 'an idea, behaviour, or style that spreads rapidly from person to person within a culture'. Memes are self-replicating, like genes in biology; fragmentary; easy to disseminate by writing, cartoon, poster, graffiti, speech, or gesture; propagated widely by social media; defying analysis, emphasising clichés and slogans. They are integral to brand identification, and to sporting or tribal loyalties and styles (baseball caps; selfies; football colours; logos; body-piercing; tattoos; skateboards; motor bikes; slashed jeans; dark glasses; smoking; burkas; hoods; crucifixes).

They are essentially pre-literate.

Increasingly, political processes rely on memes even more than words, especially with identity politics. And you can't formulate a debate with memes, only make assertions: another tattoo, an ever-louder motorbike, one more piece of graffiti.

Scott Morrison (like Tony Abbott) largely communicates in memes and the repetiton of clichés ('fair dinkum'; 'congestion buster'; 'hard-working Australian families'). If a gold medal was awarded for avoiding direct answers to questions, Morrison would be an obvious winner.

Even worse has been his use of exaggeration, fear-mongering, half-truths, and lies. A cynical defence was offered by Morrison that Shorten's campaign in 2016 against possible changes to Medicare, dubbed 'Mediscare', meant that truth was no longer a tradable commodity in election campaigns. Like Trump, Morrison could say anything—and get away with it.

Also like Trump, Morrison's limited vocabulary may have helped him to win the election. He kept repeating the words or phrases: 'work', 'home', 'family', 'How good is …?', 'humble', 'quiet', 'reward', 'amazing', 'the greatest country on earth', 'The Sharks'. Anybody exposed to a barrage of advertising slogans 24/7, as they say, may now be desensitised to nuanced argument.

One could be confident that among the words that Morrison would never use in an election campaign are 'environment', 'global', 'planetary', 'nature', 'creativity', 'imagination', 'understanding', 'explanation', 'science', 'research', 'evidence', 'books', 'art', 'music', 'beauty', and 'moral leadership'. He has his own interpretation of 'freedom'.

Again, like Trump, Morrison seems to be completely lacking in curiosity. On issues raised with him, he either knows the answers already, or has no desire to hear the case for and against a proposition. He was caught on television in 2018 on a drought-ravaged farm, refusing to link the drought with climate change:

> It's not a debate I've participated a lot in in the past, because I'm practically interested in the policies that will address what is going on here right and now. I'm interested in getting people's electricity prices down, and I'm not terribly interested in engaging in those sorts of debates at this point.

The 2019 campaign was infantilised by Morrison's televangelism. But it clearly worked.

I was concerned about Morrison's adoption of some of the Trump techniques—not just the baseball cap, but also the use or misuse of language, relying on photo ops, sound bites, and mantras. The language was simple, stripped of meaning, but endlessly repeated, again and again, over and over, on and on. At least Morrison generously let Clive Palmer have the slogan 'Make Australia Great' (why not 'Again'?).

As a religious person (unlike Trump), he has an oddly casual approach to the truth of a proposition. Morrison is essentially a salesman, a Willy Loman. In an age of retail politics, the fundamental issue for him is, 'Will it sell?' He was essentially selling a single product: short-term self-interest.

His ludicrous journey to Christmas Island, with a bevy of media, in March 2019 was based on the proposition that passage of the Medevac legislation would result in an upsurge of refugees arriving by boat. When that didn't happen, the next we heard of Christmas Island was in the April 2019 budget, with costings for closing down the detention centre. In just three weeks, Christmas Island had gone from being essential to becoming pointless.

First Dog on the Moon (Andrew Marlton) scored a direct hit with his cartoon in *The Guardian Weekly* on 24 May 2019: 'A mid level marketing manager who was as surprised as anyone that he won! Morrison had no front bench, no policies, no record to stand on, a government defined by five years of cruelty, chaos and grift and he beat the Labor Party BY HIMSELF! ONE GRINNING PENTECOSTAL IN A BASEBALL CAP AND HE BEAT THE ALP AND THE UNION MOVEMENT ALL BY HIMSELF'. (In capitals for emphasis, as if any was needed.)

History

I am old enough to have a vivid recall of the Great Depression, World War II, the personalities of Stalin, Hitler, Churchill,

Roosevelt, Gandhi, Mao, the Atomic Age, the Cold War, the end of colonialism, the Korean and Vietnam Wars, and the collapse of the Soviet Union. Competing ideologies were central during this long, transforming era—and so political outcomes were essential for survival, and personal commitment to public life seemed to be compelling.

However, for later generations, most of the names mentioned above mean nothing. Most Australians have lived through the digital age, with 29 years of unbroken economic growth (until the coronavirus pandemic struck), when all values have been seen as economic, the horizon limited, the time scale immediate, all outcomes personal, and—continuing wars in the Middle East notwithstanding—foreign issues mostly related to travel and cultural consumption. Access to data has been easy—absolutely pain-free—and you can choose your own facts.

When politicians harangue their audiences, faithfully following their scripts, they often invoke great names from the past, assuming that listeners have a shared memory that bears a powerful emotional weight. This is very dubious. Robert Menzies retired as prime minister in 1966. It is unlikely that anyone under the age of 60, even avid consumers of books and the web, will feel much stirring of the blood when Scott Morrison says, 'The Liberals remain the party of Menzies.'

Back in 2007, I was invited to address drama students at a university in Melbourne before they went to see *Keating!*, the musical written by Casey Bennetto, because they wanted to know who Paul Keating was. Most of the students would have been about ten when he was defeated in 1996, so it was understandable that they were unfamiliar with him.

I said, 'We've never been very good in identifying major figures in our history. There's only one Australian from the nineteenth century whose name everybody knows—an iconic figure celebrated in the paintings of Sidney Nolan and in films starring Mick Jagger and Heath Ledger.' Of course, I was referring to Ned Kelly. I went on, 'I'm sure you all know who I mean?' From the baffled looks on the students' faces, it was clear that they did not. Two hands went up. The owner of the first hand ventured 'Captain Cook?'; the second, 'Bob Hawke?'

On another occasion, I talked to a group of highly qualified 40-year-olds, for whom the name 'Stalin' created only the faintest flicker of recognition, and Hitler and Churchill a little more. However, the film *The Death of Stalin* may have changed this, not to mention the Churchill films starring Brian Cox and Gary Oldman.

Lack of collective or historical memory
The celebration of Australia Day on 26 January is a contentious issue in the culture wars. Even its staunchest defenders are not sure what it commemorates: some think it is James Cook's landing in 1770; others, correctly, that it is the arrival of the First Fleet under Arthur Phillip in 1788.

'Australia Day' is a very New South Wales event. In any case, it was not a cause for celebration by Aboriginal and Torres Strait Islander people, who lost exclusive rights to their homeland, and then had to endure the introduction of disease, the devastation of massacres, and the destruction of much of their culture.

Scott Morrison's federal electorate is named Cook. It was in Botany Bay, Kurnell, part of his modern-day electorate, that Lieutenant James Cook, after firing three shots and wounding local Aboriginal people, made landfall with his party on the east coast of Australia on 29 April 1770, and raised the Union Flag.

After charting the whole eastern coastline, he raised the Union Flag again on 22 August 1770, at Possession Island in the Torres Strait, and claimed possession of the whole eastern cost, which he named New South Wales, for King George III. Morrison provided

$6.7 million for a 250th anniversary 're-enactment' (his word) by a replica of the *Endeavour* of Cook's circumnavigation of Australia. However, there were two significant problems with this.

First, Cook never circumnavigated the continent. He made three great voyages in the Pacific, but observed Australia (then known as New Holland) only on the first. He arrived from New Zealand, charted and explored the east coast of Australia, and then sailed back home via the East Indies (now Indonesia).

Second, the year 2020 marked the 232nd anniversary of Arthur Phillip's arrival.

Presumably, Morrison's advisors did not draw attention to his slips.

However, James Cook and Scott Morrison had one factor in common: both would have regretted visiting Hawaii.

As it turned out, the social-distancing measures introduced in the wake of the coronavirus outbreak put paid to the projected stunt.

CHAPTER TEN

Australian Exceptionalism

In the United States, people often speak and write about 'American exceptionalism' and about 'one nation under God'. 'Prosperity Christianity', while stopping short of insisting that God is American, sees the hand of God in establishing United States hegemony, and rejects any argument to the contrary.

In practice, this means that Americans deeply resent any criticism or any comparison that suggests other nations may have superior political, health, education, race relations, gun-control laws, or criminal justice systems, and this may extend to cuisine, or film, or airline safety, or even construction for public works.

There is a different kind of 'Australian exceptionalism'—an important but unrecognised feature of this country. It is the central element in the political paralysis whereby both government and opposition are on a virtual unity ticket in congratulating themselves on the country's heroic qualities, while failing to tackle issues that they fear might be unpopular or arouse the fury of powerful vested interests.

Indeed, a code of silence (*omertà*) has been adopted in Australian society that the Sicilian Mafia would feel very comfortable with. On many issues, discussion is suppressed or minimised by the major parties, the media, and the community at large. Some issues we prefer to ignore or treat preferentially.

Australian exceptionalism has preoccupied me for some years, and brought me close to an unlikely ally, Malcolm Fraser, who served as Liberal prime minister from 1975 to 1983, first taking office under controversial circumstances after the governor-general, John Kerr, dismissed the Whitlam government, but then winning three elections (in 1975, 1977, and 1980) before losing to Bob Hawke (in 1983).

After leaving parliament in 1983, Fraser became increasingly progressive. He invited me to join the board of CARE Australia, which he had founded in 1992, and we worked together closely. On some issues, such as the Republican referendum of 1999, he formed an unlikely alliance with Gough Whitlam, collaborated in campaigns with him, and shared the same list of enemies. He even delivered the 2012 Whitlam Oration at the Whitlam Institute in Sydney.

Courage Party, anyone?

Fraser thought that both major parties had become corrupted and timid, looking mainly for competitive political advantage and adopting a narrow focus on economics—as if humans could be defined as consumers only, as *Homo economicus*, and as if the goals of life were entirely material. We shared the view that the major parties were failing to act decisively on the significant long-term issues facing Australian society, the most obvious of which is the existential challenge of climate change.

Malcolm had resigned from the Liberal Party in 2009—despite being a life member. For myself, as a life member and twice former national president of the Labor Party, I thought that expulsion would be preferable to resignation, which to me seemed weak—just walking away.

We found we could identify more than 30 crucial issues over which both major parties had shared responses—they had been captured by special-interest groups and preoccupied with the 24-hour news cycle and the short electoral cycle. In effect, the policy guide to action of the major parties could be summarised not as 'What is right?' but 'Will it sell?'

We agreed that the Australian political system was broken, the impetus for evidence-based reform on a number of important policy issues had been lost, and trust by voters in the political process had been severely eroded. As former Labor Senate leader John Faulkner put it, 'Our democracy is drowning in distrust.'

Despite our different party affiliations, we thought the major political parties had become undemocratic, structurally moribund, heavily factionalised (especially Labor), and preoccupied with playing politics ahead of achieving good policy outcomes, and too often asserting the priorities of vested interests over national interests. Put another way, we thought that Liberal and Labor had degenerated into parties of patronage and power, rather than principle and policy, and they had become narrowly focused on winning rather than serving.

The list we drew up pointed to the major parties stifling reform on a number of fronts through their shared positions on key issues. The list of the neglect and pusillanimity by the major parties seemed endless.

We hypothesised that a new political party could emerge from disaffection with the two major parties. It would not have been a 'centre party' that explored the policy differences between the major parties—where any could be found—and then split the difference, opting for something safe in the middle, offending nobody. It would have been more radical on most issues than other parties, and dedicated, to repeat the words of Leszek Kołakowski, 'to a number of basic values, hard knowledge, and rational calculation'.

I proposed the name Courage Party—which he did not warm to, oddly.

Malcolm drew upon his formidable network to ask leading experts in foreign policy, taxation, defence, environment, science, health, education, integrity regimes, and law reform, including drug laws, if they would prepare detailed position papers, analysing the evidence, and proposing long-term policy solutions to seemingly intractable problems. They all agreed. Each of the experts was dismayed by the failure of both government and opposition to act courageously on the great issues of our time.

As he aged, Malcolm became more open, more progressive, more attracted to the universal, more outraged by opportunism, and more courageous—especially on refugees, human rights, and foreign policy. At 84, he was better than he had been at 64, and far superior to the 44-year-old prime minister he had once been. That evolution was remarkable. (The same thing happened with Tim Fischer, the former National Party leader and deputy prime minister, and I spoke of that transformation at his funeral.)

Since Malcolm's unexpected death in 2015, I have expanded the list of Australian problems that our politicians refuse to address.

Indigenous issues

The myth of 'Terra nullius' was implicit in James Cook's declaration of British sovereignty, and it applied for over 200 years, from 1788 to 1992. One result of this is that Australia has never entered into a treaty with its Indigenous people (unlike Canada, New Zealand, and even the United States).

Two major elements of the British settlement of Australia after 1788 were the convict system and the dispossession of the First Peoples. Both contributed to an authoritarian strain in the Australian system, which remained, although there was also a more liberal, open, democratic, and sometimes larrikin national narrative.

The Indigenous population in 1788 has been estimated as having been between around 300,000 and 1 million, with 750,000 the most likely estimate. Most people lived in the south-east of the continent—similar to the contemporary distribution of the population, generally.

The introduction of disease, especially smallpox, as early as 1789, decimated Indigenous communities. There were hundreds of massacres in what has been called 'the frontier wars', later a hotly contested issue in 'the culture wars'. Lyndall Ryan has estimated that between 1788 and 1930 there were 65,000 Indigenous killings in Queensland alone, with many thousands in other colonies/states. The last officially sanctioned massacre occurred in 1928 in the Northern Territory, near Coniston Station.

Aboriginal people were driven off their traditional lands, forbidden to speak Indigenous languages, and often tormented, starved, shot, or poisoned. Many children were removed from their mothers ('the stolen generation'), as official policy, until about 1970. There was no attempt to negotiate treaties with the First Nations' peoples.

As late as the sesquicentenary of British occupation in 1938, some textbooks and museums classified Aborigines as 'fauna'. This was part of what ANU anthropologist W.E.H. Stanner called 'the great Australian silence'.

When the Commonwealth of Australia was inaugurated in January 1901, the premier of New South Wales, Sir William Lyne, observed: 'Of the three great colonial possessions, Australia's lot has been the happiest. Unlike Canada and South Africa, she has not had a race problem to solve.'

In the nineteenth and much of the twentieth century, 'the passing of the Aborigines' was taken for granted. In recent decades, the authoritarian treatment of Indigenous people was justified by the explanation, 'We are doing it for their own good.' This excused the exercise of rigidity, harshness, cruelty, and even sadism in institutions such as the armed forces, churches, schools, and orphanages.

The 2016 Australian Census recorded an Indigenous population of 798,000. Aborigines and Torres Strait Islander peoples number 3.3 per cent of Australia's population, but 28 per cent of all prisoners, 13 per cent of homicide victims, and 11 per cent of those charged with homicide. Adult Aboriginal and Torres Strait Islander prisoners—as a proportion of all incarcerations—range from 9 per cent in Victoria (691 prisoners) to 84 per cent (1,477) in the Northern Territory.

The draft *Closing the Gap* report for 2020 sets the goal of 'parity of incarceration' for the year 2093. The date suggests a combination of pessimism, cynicism, realism, and black humour (the adjective is unavoidable).

'Blinding trachoma' was eliminated in Europe and the United States in the 1950s, but remains in some remote Australian Indigenous communities due to the lack of clean water, and was ignored by state and territory governments until 2008. There are many other serious problems involving Aboriginal health, especially renal failure.

Indigenous issues remain highly contentious to much of the Coalition's base, demonstrated by the conservative government's failure to endorse the 'Uluru Statement from the Heart' (May 2017).

The mythic status of Gallipoli

Alan Bond hailed the win of his yacht *Australia II* in the 1983 America's Cup as being 'the greatest Australian victory since Gallipoli'. Many Australians felt the same way.

However, it is important to resist the belief that Gallipoli is Australia's—that is, White Australia's—great creation myth, and that the ANZAC tragedy brought us together as a nation. Gallipoli was, in many ways, the unmaking of a nation.

Australia was a far more vital, optimistic, and creative place in the twenty years before 1915 than in the two decades following.

Major reforms in 1895–1915 included, for example, Federation; enabling votes for women; setting up federal institutions such as the High Court and the Commonwealth Bank; establishing the Arbitration system and national wage standards, and old age and invalid pensions; and electing the world's first Labor national governments. During this period, Australia also established its own stamps and currency, started to explore Antarctica, began the Trans-Australian railway, chose Canberra as the national capital, and fostered important developments in science and education. There was a strong sense of independence, and a recognition of Australia as a progressive social laboratory.

There were fewer comparable achievements in the period 1915–35, apart from establishing the CSIRO and the ABC, inaugurating Canberra, setting up the Commonwealth Grants Commission, and building the Sydney Harbour Bridge.

The argument that Gallipoli was central to establishing a modern, confident, innovative Australia is demonstrably false. After 1915 we became more defensive and anxious, more dependent on the British connection, more derivative, and more divided on sectarian lines. There was heroism, sacrifice, and stoicism at Gallipoli, but the invasion was deeply flawed. Ironically, the over-emphasis on Gallipoli has obscured Australia's significant role on the Western Front in the last months of the war, under the leadership of General Sir John Monash, Australia's foremost military commander.

Sectarianism—a deep Catholic v. anti-Catholic division—was a toxic element in society for about 80 years, provoked in part by the abortive Irish attempt to secure Home Rule in 1916.

Politics and law

Institutionally, Australia is special for all the wrong reasons. It is the most secretive of all Western democracies, with the highest degree of unchallenged ministerial discretion, and ministerial power often excluded from judicial review. It has a very strong executive and a very weak legislature, with party discipline in parliament almost North Korean in its rigidity—in contrast, say, to the British House of Commons.

There is no independent body, nationally, to examine corruption in public life. Ministerial and public-service offices are open to lobbyists, but exempt from public scrutiny; politicians and bureaucrats are free to change their careers to become lobbyists on behalf of vested interests—often in the very areas they had nominally been responsible for.

Australia is the only common law nation without a Bill of Rights, although the Australian Capital Territory (2004), Victoria (2006), and Queensland (2018) each have one.

Great Britain adopted its Bill of Rights in 1689; the United States in 1791 (although its controversial provision protecting the right to bear arms is often cited by opponents); Ireland in 1937 (expanded in 2018); Canada in 1960 (expanded in 1982); and New Zealand in 1990. It has to be conceded that some Bills of Rights (for example, in the USSR under Stalin, and in China) have been meaningless in practice.

Decisions on defence, foreign policy, surveillance-state policies, immigration detention, trade deals, and going to war are routinely kept secret. Australia is a relatively closed society compared, say, to the US or the UK. In 2009, Labour prime minister Gordon Brown set up an inquiry into Britain's entry into the Iraq War in 2003, chaired by Sir John Chilcot, a retired civil servant. The Chilcot Report, devastating in its findings, was not published until 2016. It is inconceivable that an Australian equivalent of the Chilcot Report could be undertaken.

Defence procurement is a black hole. Billions of dollars are wasted on virtually every significant acquisition. How many submarines does Australia need? Twelve? Why twelve? Why not ten? Or fourteen? The budget implications are enormous, and yet we are not told why decisions are made—it's just, 'Take it ...'

Trade negotiations are always opaque. Details of the Trans-Pacific Partnership, for example, were kept secret—and could only be revealed, to a limited degree, *after* Australia had signed up.

The quality of debate is irrelevant in the Australian parliament because arguments never change outcomes—with party discipline enforced on both sides of the chamber, only numbers matter, not logic. Prime ministers Thatcher, Major, Blair, May, Brown, and Johnson all lost some votes in the House of Commons—and carried on. It could not happen in Australia's House of Representatives because party discipline is so rigid.

Australia's draconian *Anti-Terrorism Act No. 2, 2005*, passed by the Senate after less than six hours' debate, is harsher than comparable legislation in the US and UK: it imposes heavy penalties for committing, participating in, recruiting, supporting, advocating, or justifying acts of terror. What about analysing terrorism? Conducting research? Attempting to explain or understand it? Where is the line to be drawn expressing support for a cause and encouraging physical attack? There are legitimate fears that the laws might inhibit research or reportage. Children can be held in secret preventative detention—and it is an offence (with a maximum penalty of five years' imprisonment) for a parent to tell a spouse that their child is being held. Jesus, as a person of Middle Eastern appearance, might well have been detained under the Act.

Labor's moral nadir was its support of the *Australian Border Force Act 2015*, which provided jail terms for whistleblowers—doctors, nurses, social workers, teachers—who observed cases of neglect or abuse of refugees held as prisoners on Nauru or Manus Island and went public with their concerns. Labor abjectly voted with the Abbott government to pass the legislation. In practice, the legislation has not silenced whistleblowers who have horror stories to report, and it is hard to imagine an Australian jury convicting them, or a judge sending them to jail.

The trial of lawyer Bernard Collaery and his co-defendant—designated by the Kafkaesque identifier, 'Witness K'—for blowing the whistle on Australia's unconscionable deception of Timor l'Este in 2004 over its rights to oil and gas in the Timor Sea was held in secret in Canberra and cannot be reported on.

Refugees by sea: cruelty as a political asset

In January 2017, prime minister Malcolm Turnbull and newly inaugurated US president Donald Trump had a telephone conversation about the deal that had earlier been secured with president Barack Obama to transfer refugees from Manus Island and Nauru to the United States.

During the conversation—the transcript of which was later leaked—Trump said that Obama's Australian deal was 'the dumbest I have ever come across', and asked Turnbull why the US should take in criminals that Australia couldn't handle. Turnbull explained patiently that these people were not criminals and that they could make very fine citizens. But Trump, not exactly noted for his finesse in matters of law, said that they must have been convicted of something, or else they would not have been locked up.

Not at all, Turnbull assured him. They had broken no law, but they had attempted to arrive by boat. If they had come by plane, they would not have been in detention. 'We said if you try to come to Australia by boat, even if we think you are the best person in the world, even if you are a Nobel Prize–winning genius, we will not let you in …'

At this point, Trump gave up.

He commented to Turnbull, 'You are worse than I am.'

Later, he observed, 'What is this thing with boats? Why do you discriminate against boats?'

Then Trump got it, and the penny dropped: 'No, I know,' he said. 'They come from certain regions. I get it.'

The US arrangement went ahead in a tokenistic way, but the issue remains: the detention of asylum seekers, the abusive neglect and deprivation of children in detention, and the brutal treatment of people in offshore processing centres, mean that Australia has committed gross, unconscionable breaches of our obligations under international law.

Social cohesion and class

There is a general refusal to acknowledge that Australia has a highly stratified quality of life, as demonstrated by statistics for socioeconomic status—life expectancy, education, economic autonomy, job security, and free time.

The country's health problems are considerable: for example, it ranks no. 5 in the world for obesity; neurological disease (including depression and substance abuse) is estimated to cost the nation $74 billion annually; and there are far more deaths from the use of legal drugs than from illegal ones.

In the meantime, there are continuing high levels of domestic violence and sexual abuse.

Is Australia a classless society? Certainly not, and yet to even suggest examining social stratification provokes the accusation of trying to generate class warfare. The implications for life expectancy are very significant—but 'class' is a word that cannot be mentioned in election cycles. Obesity and smoking are strongly related to class. There are significant differences between life expectancy in Australian electorates—highest in wealthy inner-urban seats, and lower in regional and remote areas.

Media and the Arts

Australia has the world's highest concentration of media ownership and is where the Murdoch newspaper empire has its largest footprint. At the same time, the ABC is under increasing financial pressure and government attack, despite its high level of support in the community and its important role in reporting on crises. Of 33 OECD nations, Australia ranks no. 26 in its publicly funded support for the Arts—music, literature, theatre, and film.

A good society is one where engagement with creative expression and the Arts—in all its forms—are at the heart of its cultural life. Failing to value and celebrate artistic endeavour and multicultural diversity in our literature, performing, media and visual arts and crafts is to overlook its vital contribution to our culture and community.

Gambling and debt

Australia ranks no. 1 in the world for per capita expenditure on gambling, and hosts 20 per cent of the world's gaming machines. The states have become dependent on gambling revenue, and the industry also provides significant employment (including for retiring politicians and their staffers); consequently, neither major party will curb it. The combination of housing-mortgage debt and gambling debt are implicated in the fact that Australia ranks no. 2 in the world for per capita household debt.

Anomalies distorting the economy

Significant elements in Australia's economic and political evolution have been essentially based on anomalies, including trusts, tax incentives, tariffs, and subsidies. Negative gearing, for example, has made property the preferred investment, not only for the super-rich, but for middle-income families. This has contributed for decades to the real estate boom and has made negative gearing politically untouchable. It has also deprived new industries of investment.

Australia is also the only country in the world where, because of the provision of franking credits for dividends, rebates can be paid where no taxation has been collected: it is, in effect, a negative income tax.

The economic viability of rural communities and families is often structured around anomalies, driven by politics. Once created, distortionary factors are virtually impossible to eliminate because jobs and communities become dependent on them, becoming virtually welfare industries, in which the main output is not goods or services, but employment.

For almost a century in Australia, 'protection' was the major instrument for creating industries that could not survive without tariffs. In the 1980s, for example, Australia had eight motor manufacturers/assemblers, all owned overseas, at a time when the United States had only three.

Once governments provided support, it proved almost impossible to remove, unless there was bipartisan political agreement—as occurred when Hawke and Keating progressively reduced tariffs. Local motor-manufacturing, textiles, and clothing faded away over 35 years.

Today, the proposed Adani coal mine in Queensland's Galilee Basin depends on taxpayer-funded railway lines being built, any revenue to government being foregone for decades, and huge amounts of water provided without charge for 30 years.

Imagine having an aluminium smelter at A, processing bauxite that is railed and shipped from B more than 3,000 kilometres away. A is dependent on its energy supply from C, which is 500 kilometres away, involving a 25 per cent loss in the course of transmission, using brown coal, a particularly dirty fuel—creating 21 tonnes of carbon dioxide for every tonne of aluminium produced, compared to a global average of 7.5 tonnes. The smelter receives a significant public subsidy over many years, and then the owner wants to close it down because it is no longer economic.

Does any of this make sense? Establishing the smelter in A makes a great deal of sense in A, but nowhere else. In A, the community, schools, hospitals, shops, and entertainment outlets are largely dependent on the smelter. The smelter is essentially a welfare industry, and its main output is employment.

Could pineapples be grown in Tasmania, and tulips in north Queensland? Absolutely, provided there was sufficient capital to build greenhouses, and enough electricity for heating and cooling. Would it make any sense? None at all, but if a subsidy was provided it would prove difficult to remove if it would upset voters in Coalition seats.

Irrigation is another anomaly, whereby areas that cannot rely on rainfall to grow crops are given water transferred from the environment, but in times of drought cannot survive without it.

In New South Wales and Queensland, the black-coal industry uses as much water as 5 million people, and is, in effect, publicly subsidised.

Agriculture

The importation of European farming methods into Australia's huge open spaces, the destruction of forests, the dominance of monocultures, the introduction of rabbits, the early adoption of a 'knock it over, and move on' approach, and the over-reliance on chemicals and artificial fertilisers have all damaged the environment.

Nature tactfully suggests that cotton should only be grown in the tropics, north of Capricorn, where there is access to huge amounts of water. From 1983, Cubbie Station—almost 1,000 kilometres further south near the Queensland/New South Wales border, and at 93,000 hectares, the largest irrigation project in the Southern Hemisphere—grew cotton and took between 200,000 and 500,000 megalitres annually from the Culgoa River, a tributary of the Darling. This is unsupportable during drought years.

The powerful cotton lobby has close ties to the National Party.

'Virtual water' (or 'embedded water') is the total amount of water used to produce a good or service, and the estimates include costs of producing, processing, packing, and shipping. But the figures are startling. Cotton is estimated to require 11,000 litres of water for each kilogram produced; beef, 15,000; sheep-meat, 10,500; pork, 6,000; rice, 2,500; wheat, 1,500; milk, 1,000; apples, 800; wine, 500; and beer, 300. At the extremes are coffee beans (21,000 litres per kilogram) and apples (70 for each one.)

In *A Water Story: learning from the past, planning for the future* (CSIRO Publishing, 2020) water management scientist Professor Geoff Beeson cites estimates that Australia 'exports' on average more than 70,000 gigalitres of virtual water per annum.

Growth and environment

In Australia, as elsewhere, growth is regarded as an end in itself (although biology and the environment raise serious doubts about this). The country has unquestioningly adopted the worst elements of US corporate practice, including treating the environment as the enemy of 'progress'.

As David Pilling pointed out in *The Growth Delusion* (2018), in measuring economic growth, 'heroin consumption and prostitution are worth more than volunteer work or public services'.

Is all growth good? Cancer—a rapid and destructive growth that creates cellular life but destroys the host—is a striking challenge to this idea.

As we've seen, Australia ranks no. 1 in the world for per capita expenditure on gambling. Should we be aiming for more? The proceeds of gambling (the receipts actually recorded, anyway) contribute to GDP, and governments love the revenue produced by casinos. If money-laundering was confined to casinos, governments might even welcome the prospect of a higher tax return.

We rank no. 2, it appears, in drug consumption, including of opioids and pharmaceuticals. Should we aim for no. 1?

As incarceration rates go up and more private prisons are built, the GDP surges. The liquor industry is another boom area. Should we aim for more prisons and a greater consumption of alcohol?

There has been a virtual declaration of war on the environment in recent government actions, as if preserving the biota and the goals of clean air, clean water, and clean soil would destroy prospects for employment and higher consumption.

The environment is the totality of all there is in our world— the planet itself, soil, air, water, biota, and minerals. Environmental concerns cannot be regarded as residual matters after the economy has had its whack.

As Tim Wirth, a former US senator, expressed it: 'Stated in the jargon of the business world, the economy is a wholly owned subsidiary of the environment. All economic activity is dependent on the environment and its underlying resource base. When the environment is finally forced to file for bankruptcy because its resource base has been polluted, degraded, dissipated, or irretrievably compromised, then the economy goes down to bankruptcy with it.'

Education, class, and segmentation

Australia's education system has an unusually high degree of segmentation, demonstrating deep class division, with schools as a sorting device. Australia has the fourth most socially segregated school system in the OECD, behind the Netherlands, Ireland, and Chile. It also appears to be the only country in the world where private schools receive taxpayer-funded subsidies.

In 2019, Australia had 3,948,811 pupils in schools. Just

65.7 per cent were enrolled in government schools, 19.5 per cent in the Catholic system, and 14.8 per cent in independent schools.

In the United Kingdom, which was once the model for our educational practice, government schools cater for 93.5 per cent of students. In Canada and New Zealand, the percentages are even higher. In the United States, the figure is 87 per cent.

No issue is more highly charged politically (even including climate change) than public funding for private schools—Australia's experiment in transferring taxation benefits upwards. This funding creates and reinforces a powerful vested interest and is highly significant in bonding political support.

And yet, despite high levels of expenditure on education, Australia is steadily slipping in the Program for International Student Assessment (PISA) rankings for mathematics, reading, and science.

An ABC investigative report in 2017 found that, over a five-year period, 'Wesley College, Haileybury College, and Caulfield Grammar in Melbourne, together with Knox Grammar in Sydney, spent $402 million [on capital improvements]. They teach fewer than 13,000 students. The poorest 1,800 schools spent less than $370 million. They teach 107,000 students.'

If an eccentric Australian treasurer were to offer financial advantages to purchasers of Maseratis and Bentleys (which is not too far from an idea floated recently in Papua-New Guinea), such advantages, once in the system, would be very difficult to remove, and energetic lobbying would operate at the highest levels.

Some independent and high-end Catholic schools have become international players with overseas campuses—a demonstration of education having become an industry, involving some high-risk investments.

The strength of a large, comprehensive state system is that it permits/encourages diversity *inside* school and social cohesion *outside* it, rather than cohesion inside school and diversity (often harsh or fragmented) outside it.

Instead, public education in Australia has become a residual category.

Middle-class flight means that even when parents have attended state schools themselves—and could otherwise be expected to fight for public education—many go on to send their children to private schools, and then emphasise 'choice' and cross subsidies by taxpayers as their priorities.

Single-sex schools are characteristic of Muslim nations in the Middle East and Africa, and in the Catholic tradition (for example, in Ireland or Quebec). Australia's proportion of single-sex schools is one of the highest in the OECD. In Britain, the high-end private schools—Eton, Winchester, Rugby—remain strictly monosexual.

Cities and urban sprawl

Australia is the world's most urbanised nation (except for Uruguay and mini-states such as Singapore), with 64 per cent of its population living in five cities, and 90 per cent living close to the coast. This has led to it ranking fourth in the world for its dependence on motor vehicles, with its major cities occupying very large areas at a low population density. There is an abiding incapacity to grasp that the inevitable consequence of freeway construction: the creation of mega-cities and higher car-dependence, leading to more competition for urban space, traffic gridlock, and longer commuting times.

Does this mean more cars on the road? Absolutely.

While Melbourne and Sydney presently rate highly on international liveability indices, this standing is unsustainable. One of the key consequences of our intensive and prolonged urbanisation is the cost of housing.

The standard benchmark measure for affordability is the median house price divided by the national annual median household income, and affordability itself is defined as three—that is, three times or less national annual median income

In 2019—in the sixteenth annual Demographia international housing-affordability survey—all five of Australia's major housing markets were rated as 'severely unaffordable', scoring 6.9 on that rating, more than double the standard for affordability. By comparison, other unaffordable national housing markets were Canada, which scored 4.4; Singapore, 4.6; the United Kingdom, 4.6; and the United States, 3.9.

In fact, in the 2019 survey, Sydney and Melbourne were respectively ranked the third and fourth most unaffordable housing markets in the world.

So, finally, what is to be said about Australian exceptionalism? At one level, it encompasses the above list of issues that define the neglect and complacency in the Australian political system. At another, it points to the unfinished business—the agenda for change—that makes necessary a new political movement to address the tainted legacy of the major parties.

CHAPTER ELEVEN

Being Honest with Ourselves

Australian history ... does not read like history but like the
most beautiful lies.

– MARK TWAIN, *FOLLOWING THE EQUATOR* (1897)

The discussion of a number of important subjects in Australia is marked by a striking lack of candour—a hesitancy about grasping the truth. Australia is and has been a country of remarkable achievement, outstandingly successful in many ways, but we could achieve far more for ourselves and humanity generally if we were more honest about them.

I was always puzzled by the frequent assertion that we owed our political structures and robust democracy to Britain. In fact, we pioneered democratic reforms such as the secret ballot, manhood suffrage, votes for women, and the eight-hour day decades before the United Kingdom.

And yet, when I matriculated, British History and Modern History were on the syllabus, but not Australian History. I learnt a great deal about the Tudors and the Reformation, but was left to my own devices to find out about Federation, the Gold Rush, or the attempted destruction of our First Nations. It was not until 1972 that the first chair in Australian History was established, when

Manning Clark was appointed to it at the Australian National University.

We have been particularly shifty about several issues.

Race

Mark Twain, the most famous American author of his time, whose *Tom Sawyer* (1876) and *Huckleberry Finn* (1884) were international bestsellers, visited Australia in 1895 and saw things that nobody around seemed to notice, and certainly did not want to discuss. He wrote powerfully about the killing of Aboriginal people, 'the bush pudding with arsenic revenge': 'The settlers ... did not kill all the blacks, but they promptly killed enough of them to make their persons safe. From the dawn of civilization down to this day the white man has always used this very precaution.'

Twain was amazed that when he raised these issues, the reaction was not outrage but indifference.[1]

The drive to secure a White Australia was one of the driving forces of the Federation movement, and Alfred Deakin, a liberal reformer on most issues, was a zealot on race.

There was a strong white-supremacist theme in the country's initial enthusiasm for participation in World War I expressed by C.E.W. Bean, later the nation's pre-eminent war historian, that Australia was the only continent without a racial mixture. (This was a bit of a stretch, but he saw Aboriginal people as marginal, irrelevant, or headed for extinction.)[2]

At the Paris Peace Conference (1919), prime minister Billy Hughes, a Labor renegade, fought successfully against including any reference to racial equality in the treaty, and is often regarded as having been extreme on White Australia. His strident racism was deeply resented in China and Japan.

However, John Faulkner, a former senator, minister in the

Keating government, and ALP national president, points out that in New South Wales, in the federal election of December 1919, Labor went even further, pushing a racist line in a pamphlet called 'The Yellow Peril', accusing Hughes of 'Foul Lies to Expose Base Treachery ... Treachery to White Australia Unparalleled ...'

The bland assurance that 'Australia has never been racist' needs to be dispensed with.

Class

'Liberty, Fraternity and —' What was the other word? How did equality fall off the political agenda?

It's been a recurrent fantasy in modern Australia to claim that this is a classless society. It never was, although it is far less class-ridden than England, and shares this characteristic with New Zealand and Canada.

Sport, especially Australian Rules Football, has been an important social-levelling factor, enabling billionaires, professionals, tradies, and the 'unwaged', as we call them now, to share a passion—or even a meat pie—for a club. And it was often racially inclusive—although fans' and club's treatment of Indigenous players such as Nicky Winmar, Adam Goodes, Heritier Lumumba, and Eddie Betts suggests a certain fragility.

When World War II ended in 1945, the distribution of population and the relative affluence in Australia, Britain, and the US could have been represented by an equilateral triangle/pyramid, with the great majority near the base.

In 1945, Britain's Labour prime minister, Clement Attlee, campaigned for 'decent housing, decent health care, decent education, and decent pensions', using progressive taxation and other social measures to bring people out of destitution, overcoming a pervasive passivity and fatalism about what could be achieved. Sixty years later, Margaret Thatcher notwithstanding, Attlee's aims, Chifley's in Australia, and Harry Truman's in the US had largely been fulfilled.

For almost 40 years, until the 1980s, the wealth and income gap narrowed. Since then it has widened, and is now a chasm. But we don't talk about it.

In the 1990s, the ALP, the British Labour Party, and the American Democrats stopped talking about 'equality', because they wanted to make a direct appeal to the votes of middle-class 'aspirationals'. The description of Labor being a party of 'the left' became meaningless.

There seemed to be a new Beatitude: 'Blessed are the aspirationals, for they shall be rewarded, whatever the social cost.'

Aspirationals might respond to some appeals to social justice, but looked to opportunities for advancement for themselves and—in the medium term—their children. Minimum standards and opportunities for advancement were promoted over equality of outcomes.

The political class adopted, and even promoted, the myth that Australia was a classless society. Even drawing attention to significant stratifications in income, education, and health was attacked as inflammatory, inciting class warfare.

And yet it is depressingly clear that the social and economic gaps in our society are widening: the trend towards convergence has reversed.

A diagram to illustrate twenty-first century Britain, Australia, or the United States would be a somewhat elongated diamond, with the largest number of people not at the base but in the middle.

There are billionaires at the top with disproportionate influence, and an underclass of unemployed, undereducated, and ill-nourished living on the margins. But the majority is in between.

And that is where elections are won or lost in Britain, Australia, and the United States.

The French economist Thomas Piketty, in his magisterial *Capital and Ideology*, points out: 'In 1970, the average income of the poorest 50 per cent [in the United States] was $US15,200 per year per adult, and that of the richest 1 per cent was $403,000, for a ratio of 1 to 26. In 2015, the average income of the poorest 50 per cent was $16,200 and that of the richest 1 per cent was $1,305,000 for a ratio of 1 to 81.'[3]

He examines the correlation of education, income, and wealth in France, the United States, and Great Britain: 'Generally, the profile in the Democratic vote is decreasing with income ... In 2016, for the first time the top [10 per cent in income] voted 59 per cent Democratic ...'[4]

'Over the past half century, the Labour Party [in Britain], like the Democrats in the United States, has become a party of the highly educated ... In 1955, the Labour Party scored twenty-six points lower among those with college degrees compared to those without; in 2017, it scored six points higher among those with college degrees and those without.'[5]

Elections can no longer be won by marshalling the contracting proletariat. There is more emphasis on higher levels of consumption—and, in education or health care, invoking the mantra of 'choice', rather than a bottom-up approach.

Martin Luther King, Jr astutely observed of the US, 'This country has socialism for the rich and rugged individualism for the poor.'

Life expectancy in Australia for each of the 151 electoral divisions in the House of Representatives demonstrates striking variations. There are twenty electorates, all in the major capital cities, where life expectancy ranges from 84.8 to 86.3 years. The twenty electorates with the lowest life expectancy, 75.5 to 81.4 years, are mostly in rural and remote areas, with the lowest figure in Lingiari, in the Northern Territory.

The widening gap between rich and poor

Australia is more unequal in wealth than the OECD average, but less so than the US, the UK, and—surprisingly—New Zealand.

The survival of the fittest—'Social Darwinism' (a concept that Charles Darwin himself rejected)—is a powerful but unspoken force in Australian society, and drives our political life.

The widening economic gap has had a heavy social impact on the underclass. Unequal access to medical services, and social breakdown exacerbated by poverty — which contributes to family breakdown, family violence, drug, alcohol, and gambling addictions, road trauma, and suicide—are significant factors in shortening life expectancy. We are too polite to refer to these issues in our public discourse.

Piketty argues that the widening gap is not a natural consequence of the evolution of the economy, but a political artefact: not an accident, but built into policy decisions about taxation, education, and health.

According to a report released by the Australian Bureau of Statistics in November 2019, the top 20 per cent of Australian households own 63 per cent of all the wealth, and the lowest 20 per cent have just 1 per cent. The average household wealth of the top 20 per cent is $2.9 million, five times more than the middle 20 per cent ($570,000), and almost 100 times that of the lowest 20 per cent ($30,000).

Australia's richest 1 per cent now has twice the wealth of the bottom 50 per cent. (The 230,000 richest now hold $2.3 trillion.)

Of course, relying on averages, while very useful tools for international or regional comparisons, can be misleading, since they emphasise the mean and discount the extremities. Statisticians sometimes joke (it does happen) that the average human has one breast and one testicle.

The relationship between wealth and class in Australia is not at all clear-cut. Alan Bond, who provided the land for the courageously named eponymous university on the Gold Coast, Queensland, was hard to identify as being from 'the top end of town'. Australia's richest prime minister, Malcolm Turnbull, was twice disposed of by his party.

The *Financial Review Rich List*, of 200 individuals or families, published in April 2020, includes 91 billionaires. (The poorest of the 200 is worth $472 million.)

Australia's *Rich List* – Top 10			
1.	Anthony Pratt	$15.57bn	manufacturing
2.	Gina Rinehart	$13.81bn	mining
3.	Harry Triguboff	$13.54bn	property
4.	Hui Wing Mau	$10.39bn	property
5.	Scott Farquhar	$9.75bn	technology
6.	Mike Cannon-Brookes	$9.63bn	technology
7.	Frank Lowy	$8.56bn	property
8.	Andrew Forrest	$7.99bn	mining
9.	Ivan Glasenberg	$7.17bn	mining
10.	John Gandel	$6.6bn	retailing

It is significant how many of the ten have a refugee background and became rich through sheer, unrelenting effort. The list demonstrates a degree of openness in Australian life.

Andrew 'Twiggy' Forrest, the principal of Fortescue Metals, is the only one of the ten from an Anglo establishment family. Educated at the University of Western Australia, well connected politically, he is a significant philanthropist.

Mike Cannon-Brookes and Scott Farquhar, both Australian-born and graduates of the University of New South Wales, are co-founders of Atlassian, an innovative software company with global reach. Cannon-Brookes is an advocate for stronger climate-change action.

Gina Rinehart is a second-generation success story: her father, Lang Hancock, made a fortune by recognising where huge amounts of iron ore would be found in the Pilbara. She is a major donor to the National Party of Australia, and a funder of climate-change denialists.

Harry Triguboff, born in China to Russian-Jewish refugee parents, migrated to Australia in 1947. He studied textiles at Leeds University and became an Australian citizen in 1961. Known as 'High-rise Harry', his sole interest was in real estate, and he became a spectacular beneficiary of negative gearing.

Sir Frank Lowy was born in Slovakia, lived in Israel, migrated to Australia in 1952, co-founded the Westfield chain of shopping centres, and has retired to Tel Aviv. A major philanthropist, he founded the Lowy Institute and had a passionate interest in soccer, serving on the FIFA board. His knighthood was awarded by the UK government.

Anthony Pratt and John Gandel are second-generation immigrant success stories.

Pratt's father, Richard Pratt, born in Danzig, emigrated to Australia in 1938, and was a footballer and actor. He took over Visy Board, which had been founded by his father, in 1969, and over the years it became a cardboard and packaging giant. In 2009, after the death of his father, Anthony became executive chirman of the company. A graduate of Monash University, he is a passionate advocate for water and food security. His friends include Donald Trump (who opened a Pratt Industries factory in Ohio in 2019) and former Labor leader Bill Shorten. The Pratt Foundation is a major charity.

John Gandel is the son of successful Polish-Jewish emigrants who established the Sussan clothing brand. He developed shopping complexes and, with his wife, has been a generous philanthropist, especially in medicine.

Hui Wing Mao (Hu Rongmao), born in China, began investing in Australian property in the 1990s, became an Australian citizen, graduated from the University of Adelaide, was (briefly) a generous donor to the ALP, and is a major property developer in Hong Kong.

Ivan Glasenberg, born in South Africa of Lithuanian-Jewish parentage, is the chief executive officer and part-owner of Glencore, the minerals conglomerate, and has, or had, South African, Australian, and Swiss nationality.

Tolerance and pluralism? Cohesive and convergent?

Do we want to be tolerant and pluralistic? Or cohesive and convergent? In a multicultural society, is it possible to be both? One can see the case for each model, but, taken to the ultimate, the implications of both are horrifying.

Tolerance and pluralism, taken to their limits, could lead to a breakdown in shared experience, and the weakening of a common language and shared values.

If we have a commitment to free speech and diversity of opinion, can we stop people from expressing a view that we find incomprehensible? Can we limit the right to free speech and to the holding of divergent opinions, however repugnant to us, that do not impinge directly on our lives?

The goal of cohesion and convergence is admirable, especially if it is based on co-operation, collaboration, and a generous inclusiveness, and applied flexibly. But it can become rigid, dogmatic, and authoritarian—and, if taken to extremes, xenophobic and punitive: *We're all Australians around here, we all speak the same language, and people who live here must conform to uniform values —differences in football codes, perhaps, excepted.*

Cruelty and injustice

We have conspicuously failed to address the real-life and moral implications of the treatment of our First Peoples and asylum seekers.

Our prosperity is partly based on Indigenous dispossession. There should be a formal recognition of First Peoples in the Constitution, the 'Uluru Statement from the Heart' (2017) should be adopted, and there should be a formal mechanism for Indigenous reporting to the parliament.

In the deadly resort to blaming the victim, refugees by sea are seen as a challenge to the job security of workers, and their misery is treated as an illegitimate claim on our compassion, welfare systems, educational services, and national security. In effect, their weakness undermines our strength. The horrors they have escaped subvert our whole concept of values and morality—where compassion is derided as foolish, and moral values are dismissed as irrelevant.

For the most cynical of reasons, refugees have been treated as faceless, nameless, and voiceless. Politicians and bureaucrats who administer and preside over the system are determined to deny refugees the attributes of humanity, convinced that once a refugee is identified as a *person*, their punitive political cause is lost.

Reforming the Constitution—and becoming a republic

Under the Australian Constitution, there are only two pre-requisites for our head of state, following the British *Act of Settlement* (1701): the King or Queen must not be a Catholic, and must be a descendent of the Electress Sophia of Hanover. So far, we have met these conditions. Is that enough for the future?

Robert French, later chief justice of the High Court, commented in May 2008:

It is unacceptable in contemporary Australia that the legal head of the Australian state, under present constitutional arrangements, can never be chosen by the people or their representatives, cannot be other than a member of the Anglican Church, can never be other than British and can never be an Indigenous person.

I find it hard to improve on that.

Aboriginal Reconciliation and the Republic are inextricably linked. The monarchist cause is essentially the last expression of White Australia, its rhetoric, culture, politics, and the habit of deference. It is a static, essentially nostalgic, position in a society that, although dynamic in some ways, is uncertain how to express itself. It is the politics of amnesia.

The republican cause is essentially multicultural, pluralistic, independent, and irreverent—in a word, Australian. However, after two decades of only muffled (even muddled) debate, the cause does not excite mass support, for example in traditional blue-collar Labor electorates.

The ANU's polling on political trends from 1987 to 2019 indicates a steady decline in support for a republic—from 66 per cent in 1999 (although the referendum was lost) to 49 per cent in 2019, with support for the Queen flatlining at 43 per cent. Since both major parties have been Trappist-like on the subject for twenty years, and News Corp has shifted from strong support to strong hostility, support for a republic is higher than I would have expected.

The strongest reason for change is that it will force us to have a serious debate about who we are and what we want to be: to come out of the closet, instead of pretending.

The great lawyer, politician, and jurist Henry Bournes Higgins (1851–1929) argued that the 1901 Constitution was an anachronism from the outset. In practical terms, it has never operated as written.

There has been a fundamental cleavage between 'constitutional practice' as it has evolved, with a little help from the High Court, and the 'big C' Constitution.

At one time, I used to ask defenders of the constitutional status quo if they were prepared to assert in public that they believed it was right:

- for the executive power of the Commonwealth to be vested in the Queen (s. 68);
- for the Queen to retain the power to veto legislation (s. 59);
- for the governor-general to be commander-in-chief of our armed forces (s. 68);
- for the constitution not to contain any express commitment to democracy or democratic practice, or to responsible government;
- for there to be no provision for a prime minister and cabinet, or an opposition;
- for us to be best described not as Australian nationals but as 'subjects of the Queen' (s. 127); and
- until 1967, for there to have been no reference to our first inhabitants, except that they were not to be counted in the Census.

For 120 years we have said, 'Well, of course the Constitution doesn't mean what it says.' And I never had any takers to my challenge.

I resisted Kevin Rudd's proposal—acquiesced in by Malcolm Turnbull—that discussion of Australia's transition to a republic should be postponed until after the lifetime of Queen Elizabeth I of Australia (and II of England): that is, that it should be determined by an external event, over which we would have no control and that was likely to be long delayed. This is pragmatic but unprincipled.

Would United Kingdom citizens be attracted to an Australian-based monarchy if King Charles III, an old Geelong Grammarian, relocated here?

I concede that as the United States is the republican model we are most familiar with, it looks very unattractive under Donald Trump. But the appeal of Prince Andrew and his family should not dissuade us from looking at other models, particularly Ireland's.

Reforming the parliament

The parliament needs serious reform, and the electorate, passive though it is, deserves it. Voters are shocked to recognise how short our sitting periods are, with an average of 67 days for the House of Representatives since Federation. Questions—both Questions without Notice (currently a pointless theatrical charade in the House) and Questions on Notice (in written form)—should be serious requests for information, deserving serious, thoughtful replies.

And both government and opposition must be prepared to have serious debates on serious subjects, notably refugees, that have the potential to embarrass both sides. And expertise should be used.

When Melissa Parke, experienced in dealing with human-rights and refugee issues from her work as a lawyer with UN agencies, was the Labor MP for Fremantle (2007–2016), she was warned by elements in the leadership group of her own party not to rock the boat and raise moral concerns about the ill-treatment of asylum seekers. This example could be multiplied several times within the two major parties.

There is an overwhelming case for setting up an Independent Commission Against Corruption (ICAC) that would report directly to the Commonwealth parliament—as does the auditor-general—to disclose serious ethical breaches and corruption, and to enable sanctions to be imposed. The New South Wales ICAC has sweeping powers, and its reports have led to trials, convictions, and jail terms for former ministers, which is precisely why politicians are reluctant to create a federal equivalent. Only an ICAC can scrutinise the role of lobbyists (who protect themselves by invoking the 'commercial-in-confidence' formula) and the operation of ministerial advisors (who operate behind a firewall of official unaccountability).

Political parties
Informed citizens should consider these questions and provide appropriate answers:

- Is the current political system marked by a deep community disillusion?
- Are the hegemonic parties essentially 'shelf companies', small, closed, oligarchic, run by factional operatives, devoted to retail politics and a spoils-distribution system, deeply influenced by lobbyists and vested interests?
- Are the hegemonic parties incapable of reforming themselves?
- Have they failed to produce courageous policies to tackle many current challenges—on climate change, refugees, resisting fundamentalism, promoting open processes and the rule of law, Indigenous recognition, protecting the planet, and a post-carbon economy?
- If the consistent answer to these questions is 'Yes', is there a moral imperative by citizens to take action?

Excessive deference to our allies

If there is a war in which our allies are committed, it is certain that Australia will be there. Sending military personnel is seen as an insurance premium that we must pay. In 1956, we backed Britain and France in invading Egypt to reclaim the Suez Canal, although the Americans disagreed. We joined in the Vietnam War, together with the New Zealanders, but the British did not. Our enemies won, and are now our valued trading partners, and Vietnam became a favoured travel destination. Why were we there? Few can remember, but surely there must have been a reason?

We were in Iraq and Afghanistan for years, but little was achieved except encouraging jihadism, reviving the Taliban, and fracturing our own Muslim community.

It is not clear what our $80 billion, and counting, submarine project for the period 2030–50 is all about. The greatest potential threats to Australia in the current era of the coronavirus are further pandemics; the loss of demand for minerals; the loss of foreign students, given that our universities are significantly dependent on them; and the loss of tourists. It is not clear what role submarines will play. But presumably the submarine project is essentially an attempt to persuade the Americans—assuming that they are listening—that we are serious players in the Pacific.

If we faced the truth about ourselves, we could recognise that Australia has impressive human resources, not just stuff we dig up, and that we are already world leaders in medicine, fundamental research, and education, and could be a leader in the global transition to a post-carbon economy.

CHAPTER TWELVE

The Corona Revolution

Every pandemic has devastated communities, caused premature death and long periods of debility, and stretched medical resources and economies to the limit. The influenza pandemic of 1918–19 killed 50 million people, more than the number of deaths in World War I. However, the global impact of the novel coronavirus (COVID-19) has been unprecedented since then. There were 555 cases reported globally by 22 January 2020. Within 46 days, by 6 March, this had risen to 100,000. Less than six months later, by 31 August, over 21 million cases—and over 845,000 deaths—had been reported. No one could predict its ultimate toll.

The coronavirus transformed the interconnected global community and become the daily preoccupation of billions, even though the total number of confirmed cases was small compared to malaria, with 228 million cases reported worldwide in 2018, 93 per cent of them in Africa. Malaria, over millennia, has probably killed more than any other disease, even the plague.

However, the health and economic implications of climate change are likely to be greater than COVID-19 by an order of magnitude if the trajectory towards 1.5–2.0C warming by 2030 proves irreversible.

The worldwide spread of COVID-19

The detection, identification, and global spread of COVID-19 has occurred with extraordinary speed. A novel viral disease, related to SARS (Severe Acute Respiratory Syndrome) and to the common cold, it was first reported in Wuhan, Hubei Province, in China on 1 December 2019, and originally named SARS-CoV-2.[1] (It may have been detected, but unreported, as early as August 2019.)

Warnings about its impact by Li Wenliang, an ophthalmologist, and other whistleblowers, were dismissed by authorities as alarmist, and Li was threatened by police. (He died of coronavirus in February 2020.) Early cases were linked to the Huanan seafood market in Wuhan. The disease probably originated with bats, and then was passed to pangolins before jumping to humans.

On 31 December, the World Health Organization (WHO) was advised about the new disease, but its initial response was lethargic, apparently accepting Chinese assurances that human-to-human transmission was rare.

On 12 January 2020, Chinese authorities shared the genetic material of the virus for use in the development of diagnostic kits. China was still doubtful about how infectious the disease was, and the WHO was slow to challenge this.

The first recorded death, in Wuhan (9 January 2020), was followed by the first case outside China; in Thailand (13 January); then Japan, South Korea, and the United States (21 January); France (24 January); Australia and Canada (25 January); Germany (28 January); India (30 January); Italy, Spain, and the United Kingdom (31 January); Iran (19 February); and New Zealand (28 February). Indonesia confirmed its first two cases very late, on 2 March.

The WHO declared a 'global health emergency' on 31 January, named the disease 'COVID-19' on 11 February, but, although it had already spread to 110 nations, failed to declare it a 'pandemic' until 11 March.

China imposed draconian methods to contain the disease within Hubei Province, but there was professional scepticism about its reported numbers of cases and deaths. However, China was soon exporting respirators and other medical equipment to the United States, which needed them.

Women leaders were often better than their male counterparts in taking early action to combat the pandemic. They were probably more easily persuaded by the evidence, while male leaders tended to be overconfident that the fears were exaggerated. The European Centre for Disease Prevention and Control reported that countries with women in positions of leadership suffered six times fewer confirmed deaths from COVID-19 than countries with governments led by men.

The state of Kerala, in the south-west of India, was spectacularly successful in taking early action, in January, to control, test, and contain COVID-19. Kerala has a Marxist government, and the campaign was driven by its female health minister, K.K. Shailaja (always known by her initials). Kerala has India's longest average life expectancy: 77 years. With a population of 33 million, there were, by 31 August 2020, 73,855 cases and only 287 deaths (0.39 per cent).

New Zealand, Denmark, Norway, Finland, Iceland, Taiwan, and Germany, all led by women, had very low rates of infection and deaths. Hong Kong, where chief executive Carrie Lam was hanging on by a thread, also had very low figures.

Australia was also remarkably successful initially under Prime Minister Scott Morrison. The premiers of New South Wales and Queensland—both women—were very effective.

Iran soon had a higher death toll than China, but the epicentre of the disease soon moved to Europe, then to North America, and on to South America and Africa.

Australia was quick to ban flights from China (on 1 February), but slow in the case of flights from Italy, which soon became the epicentre of the coronavirus. The prime minister was tactful to not identify the United States as by far the largest source of infection, either by passengers on cruise ships or on aircraft.

The death rate was determined by several factors: age (especially those who were 80 or older); pre-existing diseases (particularly cardio-vascular and diabetes); gender (male); obesity; smoking; and the use of overcrowded public transport.

There was uncertainty about whether children were susceptible to the disease. And a vaccine was unlikely to be available for eighteen months, although more than 100 research groups, including several in Australia, were collaborating on possible models.

As at 31 August 2020, the highest number of reported deaths from COVID-19 had occurred in the United States (183,000, from just under 6 million reported cases), followed by

- Brazil: 120,828 deaths, 3,862,311 cases;
- India: 64,469 deaths, 3,621,245 cases;
- Mexico: 64,158 deaths, 595,841 cases;
- The United Kingdom: 41,586 deaths, 336,670 cases;
- Italy: 35,477 deaths, 268,218 cases;
- France: 30,611 deaths, 315,813 cases;
- Spain: 29,011 deaths, 439,286 cases;
- Peru: 28,607 deaths, 639,435 cases; and
- Iran: 21,462 deaths, 373,570 cases.

Discrepancies in the numbers of reported cases and mortality rates in India, Russia, China, South Africa, and Singapore defied credibility, suggesting that reporting systems were not consistent. Scotland had a strikingly low infection rate (0.3 per cent), but a very high mortality rate (12.2 per cent).

While some US states, especially Washington and California, took early action to contain infections, the national response was slow and is described later. Not until 10 April did the total number of tests carried out in the United States exceed those in Australia (with 8 per cent of America's population). The hospital system was stretched to the limit.

There were striking variations in the case–fatality ratio, sometimes with neighbouring states, confirming that early lockdowns were crucial. Within months, however, as governments succumbed to pressure to relax restrictions, infection rates surged and second waves took hold. Here is a selection of those ratios around the world at the time of writing:

Italy 13.2	Brazil 3.1
Lombardy 13.2	Greece 2.6
Emilia-Romagna 13.6	Norway 2.5
United Kingdom 12.4	Cuba 2.4
Scotland 12.2	South Africa 2.2
Belgium 11.6	Japan 1.9
Mexico 10.8	India 1.8
Hungary 10.0	Czech Republic 1.7
France 10.0	South Korea 1.6
Canada 7.1	Australia 2.4
Sweden 6.9	Tasmania 5.7
Spain 6.6	Victoria 2.8
Sicily 6.2	Western Australia 1.4
Ireland 6.2	New South Wales 1.3
Iran 5.7	South Australia 0.9
China 5.3	Queensland 0.5
Indonesia 4.3	Taiwan 1.4
Finland 4.3	New Zealand 1.3
Germany 3.8	Israel 0.8
United States 3.1	Hong Kong 1.8
New York 5.8	Singapore 0.05

Source: COVID-19 Dashboard by the Center for Systems Science and Engineering (CSSE) at Johns Hopkins University (JHU), accessed at coronavoirus.jhu.edu on 31 August 2020

In Brazil, Donald Trump's clone Jair Bolsonaro was dismissive of COVID-19, describing it as 'just a little cold', refused to wear a protective mask, and shrugged off the rising death toll: 'We will all die one day, so what's the problem?' Later he contracted it himself—but not acutely.

Russian president Vladimir Putin's approach was also cavalier. Russia ranks fourth in the number of reported cases, but the mortality figures are suspiciously low because deaths were only counted if there had been an autopsy.

The United Kingdom was late to respond, partly because Prime Minister Boris Johnson was preoccupied with Britain's exit from the European Union. Then there was brief consideration of a 'herd immunity' approach, which Sweden tried with mixed success.

Australia was in virtual lockdown by 13 March, with panic buying in supermarkets and fourteen-day isolation periods imposed on returning travellers, but it was ten days later before the UK followed suit—and this delay contributed to many thousands of deaths.

The British prime minister, his health secretary, and the Prince of Wales were all infected with COVID-19, and Johnson spent days in hospital in an ICU ward. After his initial shambolic response to COVID-19, he expressly repudiated Margaret Thatcher's notorious assertion that 'there's no such thing as society. There are individual men and women and there are families.' He said, 'There really is such a thing as society', and that the infection rate would only be curbed with community support for social distancing and a lockdown.

The United Kingdom's mortality rate was unusually high, just behind Italy's, close to Belgium's and France's, and much higher than the United States'.

Australia and New Zealand both had very low mortality rates: just 2.5 per cent and 1.3 per cent respectively. By 31 August, Australia had 25,746 confirmed cases and 652 deaths; New Zealand,

1,738 and 22. A high proportion of Australia's infections could be traced to cruise ships, and its deaths traced to aged-care facilities.

Adam Creighton, economics editor of *The Australian*, took a courageous view, pointing to Sweden as a success story in tackling COVID-19 by promoting herd immunity without an economic lockdown: with 40 per cent of Australia's population, its mortality rate was only 37 times greater.

In Australia, following immediately after the bushfire season of 2019–20, coronavirus was a second disaster, but was far better handled. It had to be.

By 23 March, the prime minister had warned that the emergency might continue for six months or more. Sporting events were cancelled, Australia withdrew from the projected 2020 Tokyo Olympics, and cinemas, churches, restaurants, and pubs were directed to close. States closed their borders. Foreign travel was banned.

The Commonwealth budget was postponed from May to October. The prime minister and state and territory premiers/chief ministers were working closely in a 'National Cabinet', and the federal opposition gave bipartisan support for government measures.

Despite some early fumblings at the outset, and some contradictions, Morrison became confident and assertive as the pandemic proceeded. Then it became clear that his preparedness to take bold action would only be temporary, and when the crisis was over there would be a return to partisan ideology. Greg Hunt redeemed himself somewhat as health minister. (I write this with gritted teeth, but it needs recording.) Josh Frydenberg, as treasurer, was impressive with his mastery of detail, but then, after unparalleled government intervention to support wages and the private sector, invoked the shades of Margaret Thatcher and Ronald Reagan, both of whom believed that crises should be left to the market to sort out.

Coping with previous health and economic crises

The Spanish flu

The influenza pandemic of 1918–20 was usually described as the 'Spanish flu', only because Spain had been neutral in World War I, had no censorship, and became the first country to extensively report the disease.

It could have been described as Iowa flu, or Kansas, or French, or Austrian—or even Chinese. The earliest reported outbreaks were in Iowa and in a military camp in Kansas, and may have been brought to France by the US army. Blame has also been attributed to Chinese labourers brought to Europe, via Canada, for the war.

Most deaths were due to secondary bacterial pneumonia and lung failure. It reached New Zealand in October 1918, and the first recorded Australian case was in Melbourne in January 1919. The Victorian government dithered, failing to alert the other states that a new disease was spreading and could at best be contained rather than cured. Eventually, about 15,000 Australians died, more males than females, and most in the 20–40-year age range, at a death rate of 2.7 per cent.

(It should be noted that there is a puzzling discrepancy in these statistics. In 1919, Australia's population was 5.193 million and the infection rate is often quoted as having been 40 per cent, that is, 2.077 million. However, if the mortality rate was 2.7 per cent, the death count would have been 56,000.)

New Zealand had a higher death rate than Australia, especially among Māori (about 42 per cent). Aboriginal Australians appear to have suffered an even higher death rate (perhaps 50 per cent).

The Great Depression

The Great Depression began with the Wall Street collapse in October 1929, and in many nations it lasted until the outbreak of

World War II. International trade fell by 50 per cent between 1929 and 1932. Australia was particularly hard hit, with unemployment at 29 per cent in 1932. The United States, Germany, Canada, and New Zealand were among the worst-affected nations. The Depression put Hitler in power in Germany, consolidated Stalin's rule in Russia, and created Franklin D. Roosevelt's New Deal in the US.

The Asian flu (H2N2)

The Asian flu of 1957–58 caused more than 2 million deaths worldwide, mostly in China and Latin America, with a mortality rate of 0.7 per cent. Pregnant women and children were particularly vulnerable. There were 14,000 deaths in the UK, but Australian figures are elusive. It originated in China.

The Hong Kong flu (H3N2)

The Hong Kong flu of 1968–70 left a global death toll of about 1 million, with a 0.2 per cent mortality rate. There were about 100,000 deaths in the US. It affected older people, and was linked to seasonal influenza. It may have originated in pigs. The Health Department does not report the number of Australian deaths.

HIV/AIDS

HIV/AIDS (Human immunodeficiency virus/acquired immunodeficiency syndrome) caused a massive failure of the immune system, resulting in lethal conditions such as tuberculosis, tumours, and opportunistic diseases, especially pneumonia. Its impact was relatively slow, but the death rate was high. The virus originated in Africa, probably in the Cameroons, from primates, and was communicated by direct physical contact or in blood (including transfusions). The WHO calculates that HIV/AIDS has infected 75 million people, and that 32 million have died of it, mostly in Africa. Symptoms were often not detected for years, but were exacerbated by intravenous drug use.

Australia's first recorded HIV/AIDS case was in Sydney in October 1982, with the first death in July 1983, in Melbourne. A network of interested groups generated an effective grassroots campaign to provide information about the condition, followed in 1987 by the powerful 'Grim Reaper' television advertisements, with decisive leadership by the health minister, Professor Neal Blewett, and Bill Bowtell, his senior advisor.

Australia's reaction to the HIV/AIDS crisis was in sharp contrast to the United States, where President Reagan showed great hesitancy in even referring to AIDS. This was probably a combination of his queasiness about homosexuality and concern that he might shock the Moral Majority, an important part of his cheer squad. He certainly refused to be specific about condoms, sex education, or needle exchanges. And he repeated his slogan, 'Government is not the solution to our problem, government is the problem.'

The Avian flu (H5N1)

The avian flu was detected in Hong Kong in 1997, and had a deadly effect on poultry. There was little evidence of person-to-person infection, except for a few families in Hong Kong and Vietnam. It had no direct impact on Australia, except to speed up planning against a future pandemic, if and when it came.

SARS

SARS (Severe Acute Respiratory Syndrome) was a deadly form of pneumonia, transmitted by virus, with a very high mortality rate (10 per cent) in China, Hong Kong, and Taiwan. The total number of reported cases was fewer than 10,000, and it was of relatively short duration, from November 2002 to May 2003, when it disappeared suddenly. However, it had spread to 26 countries. In Australia, six people were infected, but none died.

The Global Financial Crisis

The Global Financial Crisis of 2008–09, precipitated by a Wall Street crash and triggered by the bankruptcy of Lehman Brothers investment bank, had no direct health implications, but became a model of how Australia—or governments generally—could react to a global threat. The Rudd government decided to 'go early, go hard, go households' with a $42 billion direct-stimulus package. The opposition, led by Malcolm Turnbull, voted against the stimulus. Australia retained its AAA international credit rating and was, with Israel, Poland, and South Korea, one of only four nations to avoid a recession.

The swine flu

The swine flu pandemic of 2009–10, like the influenza pandemic of 1918–20, was classified as an H1N1 virus, but was a new strain, involving elements of bird, swine, and human flu. The first report of the disease came from Vera Cruz, Mexico, and more than 200 countries were affected. It was highly contagious, but with very low mortality. It is estimated that between 0.7 and 1.4 billion people globally contracted the disease, with perhaps 300,000 deaths—a mortality rate of 0.03 per cent. It lasted for nineteen months.

The Australian Department of Health and Ageing published a detailed report on swine flu in November 2009, in which it recorded 37,435 reported cases and 191 deaths, with a median age of 53—a mortality rate of about 0.5 per cent, far above the global average. A longer study was published in 2011. Oddly, the swine flu has fallen out of the collective memory.

Ebola

The Ebola virus caused a pandemic in West Africa in 2013–16, especially in Sierra Leone, Liberia, and Guinea, with an exceptionally high mortality rate—more than 40 per cent. Bats were the original vector of the virus, and the problem was compounded by poor sanitation and lack of medical capacity. The WHO was slow to respond, and Médicins sans Frontières took the lead in tackling the outbreak. Some health workers died in the US, Britain, and Italy. Australia watched in horror, but was not directly affected.

COVID-19, social distancing, and accepting rapid change

COVID-19 has brought about revolutionary changes in work, sport, the arts, education, health, transport, tourism, and the way we live, work, entertain ourselves, and interact with others—in weeks, not years or decades. It is also exposing class and regional differences, with the most serious impacts on the unemployed, the unskilled, and the socially isolated. Professionals can work from home: part-time cleaners and fruit pickers cannot.

Even profound policy changes seemed possible for a time. During Australia's 2019 election campaign, the education minister, Dan Tehan, dismissed a Labor proposal for free childcare for working parents as 'socialist if not communist'. Nevertheless, the Coalition government waved it through (admittedly on a temporary basis) in 2020, amid a raft of hitherto-unthinkable interventions in the operations of the economy.

The nationwide lockdowns imposed to suppress the coronavirus—and the distress and dislocation they caused—immediately raised a number of critical questions. Did we need to go to work to be employed, or to school and university to be educated? What was the future of the arts, or mass participation in sports? Was it worth thinking about a Universal Basic Income instead of scattering welfare benefits? What was the future of tourism and aviation? (The future of the cruise ship industry must have been in doubt after the *Ruby Princess* fiasco in Sydney.)

Would the operation of our courts, parliaments, prisons, and medical facilities have to change? Would older people have to wear security devices to provide alerts if they injured themselves?

What were the implications for universities and students in their last year of school? Would foreign students still be welcomed? Was it the death knell for department stores and shopping malls? Would restaurants and cafés survive?

After an abject, even frivolous, failure to take strong action to mitigate Australia's contribution to climate change and to plan for transition to a post-carbon economy, Scott Morrison reacted to the coronavirus pandemic with an unparalleled series of interventions, including a community lockdown and injecting a $214 billion stimulus (11 per cent of GDP) into the economy to provide, inter alia, wage support to stem massive job losses. This represented the greatest expansion of executive power in my lifetime, based on modelling that the government was slow to share. However, it won broad support from the community, and from the opposition and state premiers. The Business Council of Australia and the Australian Council of Trade Unions both applauded the action. An astonishing $60 billion overestimate by Treasury of the cost of the JobKeeper program reduced total outlays, but they remained massive and were expanded again in August 2020 in response to further economic damage caused by Victoria's second-wave disaster.

Australia soon became engaged in a two-front 'war' on health and the economy. The government said, in effect, *We will have no economy if we have no health. Health is the more pressing. Let us fight on that front first.* But what was the evidence that supported the health measures that were being taken? What were the competing views? Of these, we knew very little. Most important of all, we did not know what economic decisions and implications were implicit in the health choices being made.

The Bureau of Communications and Arts Research (BCAR) published an analysis showing that cultural and creative activity contributed $111.7 billion to Australia's economy in 2016–17.

The closure of theatres and concert halls during the COVID-19 lockdown had a devastating effect, with actors, musicians, dancers, and film crews typically being self-employed or classified as 'casuals'. The nature of their employment excluded most from receiving benefits from the government's bailout provisions.

Universities were also specifically excluded, and with so many staff in casual or sessional employment, the impact of this was severe. Universities lost billions of dollars in fees from overseas students. While some of the older universities had substantial reserves, the newer ones were completely dependent on Chinese and Indian students to remain solvent.

What do we know about triage and how clinical decisions are made in over-stressed hospitals and ICUs? What will be the consequential impact of any pandemic on overstretched hospitals? Will it take resources away from elective surgery and cancer treatment?

If we stayed at home, would we become less engaged with the community and more dependent on devices? Would there be more anxiety, drug use, and domestic violence? Would we be more or less obsessed with gambling? Was 'isolationism'—an interesting word—or 'patriotism' needed to protect us? Had we become a surveillance state? How many of these issues were being discussed in our parliaments? (I know the answer to the last question.)

A requirement for 'social distancing' was soon set at 4.0 square metres. Handshakes and kissing, real or in the air, were out; handwashing was in. Grandparents could not visit their grandchildren. Attendances at weddings and funerals were strictly curtailed. Pubs and clubs had to shut down.

This was a revolution. David Allen Green, an English lawyer, journalist, and blogger, commented:

Three fundamental freedoms—freedom of movement, freedom of association and freedom of worship—have all been abolished for six months by a statutory instrument which has been neither scrutinised nor voted on by members of parliament.

If it were not for this public health emergency, this situation would be the legal dream of the worst modern tyrant. Everybody under control, every social movement or association prohibited, every electronic communication subject to surveillance. This would be an unthinkable legal situation for any free society. Of course, the public health emergency takes absolute priority. But we also should not be blind to the costs.

In the United States, which by the end of March 2020 already had more reported cases of COVID-19 than any other nation, President Trump, after a fanciful period of denial and magical thinking, reluctantly sanctioned a $US2.2 trillion government rescue package to prop up a failing hospital system.

In Australia, parliament met briefly on 25 March to vote for the rescue package, and then voted to adjourn until August, until there was a realisation that it would need to return to pass more measures; it soon resumed sittings, while observing social distancing.

In Britain, the House of Commons rose a week earlier for the Easter break, but resumed in mid-April. In the United States, both the Senate and the House of Representatives remained in session. The Parliament of the European Union continued to meet, online.

There was some recognition that in difficult times there must be more democracy, not less. Beyond that lay a challenge that governments not normalise restrictions on liberty.

In Australia, the prime minister took some time to shed his tendency to bristle, and to avoid and evade, when asked difficult questions, and was reluctant to disclose the nature of the advice he received—in contrast to the state premiers and chief ministers. How much was Scott Morrison relying on modelling when, on 13 March, he said it was safe to attend a football match, and then reduced the numbers of people allowed to gather in public to 500, then 100, then ten, and then (on 28 March) to two? The government was shifty in the manipulation of the data behind its modelling. Perhaps most unsettling of all, the messages about school attendance coming from federal and state leaders were confusing and contradictory for too long. However, there were grounds for optimism.

Ideologically, it seemed that three sacred cows had been put down: that growth is an end in itself, that the budget must always be in surplus, and that a more generous unemployment benefit would encourage people to rely on welfare.

And federalism provided a useful pushback. Dan Andrews, the premier of Victoria, was simple, direct, and powerful without bloviating about the economy. Andrews and Mark McGowan (Western Australia) took the lead in pushing the federal government to take stronger action, and this early advocacy was later used against Andrews when a second wave hit Victoria badly but not the other states. State and territory leaders were effective in tackling problems and explaining what they were doing.

Victoria had consistently imposed the tightest restrictions to containing COVID-19, but Andrews came under attack in July after a security failure in the 'quarantine hotels' led to an even more serious second wave, followed by some strict suburban lockdowns in Melbourne and police guards at nine of the Housing Commission high-rise apartment blocks (described as 'vertical cruise ships').

Victoria's borders were closed for the first time since the Spanish flu pandemic in 1919, and the spirit of co-operation, welcomed in the first months of COVID-19, began to break down.

Throughout the campaign against the pandemic, the good news was that the prime minister started to rely on scientific evidence, and statistics, instead of praying for a miracle. Professor Brendan

Murphy, the chief medical officer, who became secretary of the health department in June, was the prime minister's constant companion at media conferences, although his messages were not identical and usually less ambiguous.

Was it possible that after a recovery from the coronavirus, the government might use science and technology to cut our carbon dioxide emissions?

Unfortunately, this seems very unlikely. Morrison's success with COVID-19 makes it *less* likely that he will take a similar approach with climate change, and he probably assumes that the impact of COVID-19 means that all other issues will soon be forgotten or forgiven.

It soon became evident that after a 'snap back', the culture wars would resume, with science being downgraded again, public servants back in their boxes, and a continued war of attrition waged against the ABC, which had played an outstanding role during the bushfire season and the pandemic, during which, according to surveys, it was the most trusted source of information and advice (particularly Norman Swan on COVID-19).

The *Financial Review*'s editorial sternly warned that it was time to stop discussing income inequality and to cut regulations intended to preserve the environment ('green tape') that might inhibit investment. It was strictly 'back to the future', with an insistence that ultimately all values are commercial.

The National COVID-19 Coordination Commission, appointed in April 2020, had a very narrow remit, and, judging from its website, its 'focus areas' did not include environment or climate change, even though members such as Greg Combet and David Thodey had expertise in climate change, and access to relevant information. (Combet soon withdrew.) The chair, Neville Power, is a former CEO of Fortescue Metals, and flies his own jet. Andrew Liveris, a 'consultant', comes from Dow Chemicals and is a mining enthusiast.

The commission's priority (it was hard to be sure, as the organisation didn't answer questions) seemed to be to promote natural gas—that is, methane. Its chair made an early pronouncement about a 'gas-led recovery' for Australia, after which he lapsed into silence.

It soon became clear that there was an ideological underpinning to many of the government's emerging outlays. Construction work, overwhelmingly male, was favoured, while childcare, almost exclusively female, was penalised. Casual and part-time workers in retailing, hospitality, and tourism were in jeopardy. Australians who worked in the arts and entertainment—645,000 of them—were overwhelmingly ineligible for JobKeeper payments because they were not 'employees'.

The coronavirus, the United States, and Donald Trump

It would be inconceivable to many, probably most, Americans that other countries could have a better quality of life, a longer life expectancy, superior nutrition, better access to and delivery of health services, cleaner air, better water, safer cities, and fewer gun deaths. To raise comparisons with, say, Britain, Canada, Australia, or New Zealand would evoke patriotic rage and disbelief.

As it happens, the United States only ranks no. 37 in the world for life expectancy. It does rank no. 1 in its numbers of prisoners, in absolute figures: even more than China. And in the numbers of reported infections and deaths from COVID-19. To be fair, it does rank no. 1 in recipients of Nobel Prizes.

President Trump disregarded early medical advice that COVID-19 was a potential disaster. His total preoccupation was re-election in 2020, and he had little interest in or sympathy for the victims or frontline medical workers. He seemed to characterise illness as a sign of weakness, and weaponisd the word 'sick' as a contemptuous dismissal of his enemies. It was no accident that deaths of African-Americans from coronavirus were more than twice their proportion of the population.

Despite the outstanding expertise around him, Trump's interventions on the coronavirus were chaotic. His administration had weakened the highly regarded Centers for Disease Control and Prevention, which could have led an effective early response.

On 24 January 2020, Trump insisted that the coronavirus was essentially a domestic Chinese problem, that President Xi was 'doing a very good job', being 'transparent', and that the disease had been contained. Until 10 February he praised China fifteen times.

He described fears that a pandemic would spread to the United States as 'a hoax', said it was no more serious than influenza, and that there were only a handful of cases and soon there would be none. It would disappear one day, 'like a miracle'. Then talk of it was 'scaremongering by the Democrats', and then it was 'fake news'.

After 10 February, when COVID-19 erupted in Iran and Europe, he changed his emphasis. Then it became a 'foreign virus' transmitted from China and the European Union, then the 'China virus', and then a national emergency. Trump declared himself to be 'a war president', and that he had always anticipated the pandemic. 'We acted quickly and early … We are doing a great job … the best in the world,' he said.

He boasted, 'The virus will not have a chance against us … [I have put in place] the most aggressive and comprehensive effort to confront a foreign virus in modern history.'

Rush Limbaugh assured his talkback listeners, 'Folks, it's no more serious than the common cold.'

Fox News thought that coronavirus might prove to be 'a deepstate' plot.

When Trump's acting secretary of home security, Chad Wolf, gave evidence to a Senate committee on 25 February, he asserted that the mortality rate for COVID-19 and influenza were 'about the same' (although the difference at the time was 5 per cent compared to 0.1 per cent).

Mike Pompeo, the secretary of state, described COVID-19 as 'the Wuhan coronavirus' or 'the Chinese coronavirus', using framing to shift blame and create the illusion that America was under attack from its enemies.

When journalists asked about the serious lack of test kits, Trump asserted, '... the tests are beautiful. Anybody that needs a test gets a test', directly contradicting what Vice President Mike Pence had reported the previous day.

Trump has admitted to having phobias about sharks and germs, and this may account for his uneasiness about directly addressing the clinical issues involved. Before he was president, he had proposed that Americans abandon handshaking and adopt Japanese-style greetings. It may also explain why none of the Trump hotels were made available to travellers who needed to be isolated for a fortnight after foreign travel.

Trump mentioned that his uncle (an engineer) had taught at MIT, so that he knew a great deal about science himself. However, he made a rare admission: he did not know that people had died of influenza a century before (although his grandfather had been a victim in the 1918 pandemic). He said that doctors had asked him, 'How do you know so much about this?' He also declared that he had three 'hunches': that the mortality rate of coronavirus would be only a small fraction of 1.0 per cent; that the flu vaccine would be effective in treating it; and that the US was leading the world in its response. He was wrong about all three. Later, he opined that a malaria vaccine might do the trick, and said that he expected the US to be back in business by Easter. He was wrong about those, too. He briefly recommended bleach to kill the virus. Then he indicated that he was taking doses of hydroxychloroquine, an anti-malarial drug (at least he said he was, but one can never be sure), despite concerns that it could have adverse effects on overweight or cardiac-challenged users.

At a press conference, Trump was asked, 'What metrics will you rely on in making decisions about COVID-19?' He pointed to his head and said, 'The metrics [are] right here. That's my metrics.'

The central theme of his media appearances was his relentless self-congratulation about what a wonderful job he was doing, during which he showed a total empathy deficit, with barely a word for the victims or their families.

Trump was incapable of an apology (as in, 'I made a mistake') or taking responsibility for any policy failure: someone else was always to blame. Among his targets were Barack Obama and the World Health Organisation.

Attitudes to the coronavirus indicated deep divisions on party lines. In March, a Gallup poll found that 73 per cent of Democrats, but only 43 per cent of Republicans, were concerned that the coronavirus might affect their families. States with the highest number of cases were those with big cities, and on the coastline. All but one of New York, New Jersey, Michigan, Massachusetts, Pennsylvania, California, and Illinois had Democrat governors. (Louisiana, Florida, Texas, and Arizona also with a high incidence, were in Republican control.)

California, with 40 million people, ranked no. 1 in US states for its number of infections (516,000) by 3 August, and no. 3 in the number of deaths, but its mortality rate was 28th.

Eleven Republican state governors were very reluctant to adopt lockdown measures and social distancing. The governor of Florida acted, not because of the medical evidence, but only after observing President Trump's 'changed demeanour' on television.

A number of states with Republican governors abandoned restrictions early as a political gesture. Refusing to wear face masks, following the president's example, was a sign of fealty. Some of his supporters were captured on social media rebuking people who had the temerity to wear face masks on the street. There may be a US constitutional right to be infected (or shot).

Conspiracy theorists were in full flight: face masks had been designed by Satan, COVID-19 was a hoax, and/or 5G communications technology, Bill Gates, Barack Obama, and the WHO were to blame.

The state of Georgia, with a population of 10.6 million, abandoned restrictions early—and then the infection rate rose rapidly (up to 105,000 in July), with 2,922 deaths. Several Republican-led southern states that eased restrictions too early suffered surging infection rates, and had to reimpose lockdowns and restrictions.

The coronavirus pandemic resulted in increased gun purchases in the US, but this was unlikely to be a useful treatment. In Australia, the coronavirus lockdown resulted in a spike in sales of online gambling, alcohol, jigsaw puzzles, and guns.

For a long time, none of President Trump's statements reduced the zeal of his support base; polling indicated that his approval rateings were relatively stable up till May. But, quite quickly, his mishandling of the pandemic, and his lack of concern for its victims, became impossible for swinging voters to ignore. When he then responded dismissively to the police killing of African-American George Floyd, and the nationwide protests that followed, his popularity sagged noticeably, endangering not only his re-election but the Republican majority in the Senate.

CHAPTER THIRTEEN

Saving the Planet

> No man is an island entire of itself; every man is a piece of the
> continent, a part of the main. If a clod be washed away by the
> sea, Europe is the less, as well as if a promontory were, as well as
> if a manor of thy friend's or of thine own were. Any man's death
> diminishes me, because I am involved in mankind. And therefore
> never send to know for whom the bell tolls; it tolls for thee.
> – JOHN DONNE, *DEVOTIONS UPON EMERGENT OCCASIONS* (1624)

Our planet faces a series of complex and inter-related threats—not war, and probably not famine in the short term—that are likely to massively disrupt the lives of the next generation and beyond. Of three major environmental and health challenges, the most significant threat is climate change. Its major impacts may not be experienced until 2050, by when they may be catastrophic.

The major challenges stem from world population growth, climate change, and pandemics. They are all self-inflicted, all inter-related, and all could be changed, given a combination of vision, leadership, courage, and passionate engagement by informed citizens who want to preserve the world for their children and grandchildren.

World population growth and per capita resource use

In the year of my birth, 1932, the world's population was estimated as 2.1 billion people, many of them suffering very high rates of infant mortality, life expectancy below 40 years, and low levels of nutrition and resource consumption generally. The urban population comprised about 25 per cent of the total, with more people engaged in subsistence agriculture than in any other occupation.

In 2020, the world's population was about 7.8 billion, global life expectancy had increased by more than 30 years to about 71 years, infant mortality had fallen dramatically, levels of nutrition and consumption had increased exponentially (but unevenly), despite the dramatic fall in the proportion of farmers, and the urban population comprised 56 per cent of the total.

The Our World in Data (OWID) chart, based on World Bank material, is instructive. The OWID benchmark of living on less than $US1.90 ($A3.16) per day as 'extreme poverty' is not exactly lavish, and data for the nineteenth century would have been limited, crude, and highly speculative.

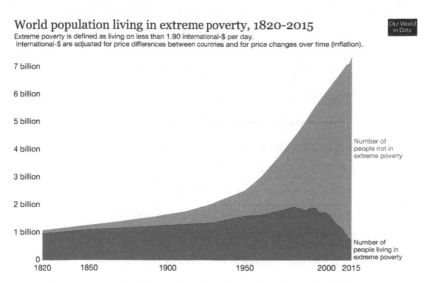

World population living in extreme poverty, 1820-2015

Extreme poverty is defined as living on less than 1.90 international-$ per day.
International-$ are adjusted for price differences between countries and for price changes over time (inflation).

Number of people not in extreme poverty

Number of people living in extreme poverty

Note: See OurWorldInData.org/extreme-history-methods for the strengths and limitations of this data and how historians arrive at these estimates.

Source: OurWorldinData.org/extreme-poverty

The World Bank estimated in 2018 that the numbers of people living in 'poverty' (distinct from the 'extreme' version above), with an almost profligate benchmark of $US7.40 per day, had been falling, slightly, since 1999. However, if East Asia (China, Japan, and South Korea) are deducted, the numbers of people in poverty are is rising, especially in sub-Saharan Africa.

The Malthusian puzzle

Thomas Robert Malthus (1766–1828), a cleric, pioneer demographer, and economist, wrote an important but pessimistic essay 'On the Principle of Population' in 1798. He argued that 'Population, when unchecked, increases in a geometrical ratio. Subsistence increases only in an arithmetical ratio.' He predicted that starvation would occur unless the fertility rate fell, which would require suppression by moral restraint.

He was writing about the population of Great Britain, not the world, but his argument could be extrapolated globally.

Malthus's thesis was taken up by the American population biologist Paul Ehrlich and his wife Anne Howland, who co-wrote *The Population Bomb* (1968), a bestseller that predicted that 'hundreds of millions' would starve to death in the 1970s, and called for 'limits to growth', just as the Club of Rome had done.

Ehrlich was soon proven wrong. Agricultural production was increasing globally at an unprecedented rate. The dramatic success of plant scientist Norman Borlaug's 'Green Revolution', which increased wheat and maize yields (for which he won the Nobel Peace Prize in 1970), improved nutrition for hundreds of millions. However, there was major collateral damage to the environment from the unsustainable use of chemical fertilisers and pesticides, and from the threats to biodiversity.

Neither Malthus, prospectively, nor Ehrlich, retrospectively, recognised the extent to which scientific agriculture could increase food production. In 1798, the world's population was probably just under one billion. There has been an almost 800 per cent increase since then—with exponential growth in agricultural production.

We now live far longer than our ancestors did. In 1798, about 90 per cent of the population would have consumed calories at or below subsistence levels; currently, 90 per cent consume at levels above it.

The increased longevity in China, India, and Africa in the past 70 years has been spectacular. Continental Africa's population was 227 million in 1950, 500 million in 1982, 1 billion in 2009, and is projected to reach 2.4 billion in 2050—a tenfold increase in a century. (Nigeria alone is estimated to reach 1 billion by 2100.)

In *Upheaval: how nations cope with crisis and change* (2019), Jared Diamond, the American geographer and historian, concludes that gross consumption of calories in 2018 was 3,200 per cent greater than it had been in 1798.

That's the optimistic view. There's an alternative view, which suggests that Malthus was wrong yesterday, but may prove to be right tomorrow (which was also said of Marx, who was another Malthusian).

In 1998, US environmentalist Bill McKibben estimated that the average American had a consumption rate of calories six times greater than the average human: the equivalent of a sperm whale, and that the biota could not support many sperm whales—or Americans.

According to Swiss-based investment bank Credit Suisse, Australia ranks no. 2 in the world, just behind Switzerland, in mean wealth per adult, and this would be reflected in our consumption rates.

India, Indonesia, Pakistan, Bangladesh, all of Africa, Brazil, most of South America, and more than 100 smaller states consume at very low levels.

If they aspired to use resources at even 20 per cent of Australia's levels—leaving aside our exceptionally high level of waste—we would need to exploit another planet (although water supply and transport might pose insurmountable problems).

I = PLOT

Consumption levels and the spatial organisation of how we live are more significant problems than population numbers themselves.

Back in 1994, the report of the House of Representatives' standing committee on long-term strategies, 'Australia's Population "Carrying Capacity": one nation—two ecologies', which was agreed unanimously, made recommendations that still have contemporary relevance. (I chaired the committee.)

The report concluded that environmental damage is not determined by numbers alone, but by population factored by per capita resource use. This was expressed in a formula proposed by Barney Foran and Doug Cocks of the CSIRO: I = PLOT, in which I stands for Impact, P for population, L for lifestyle, O for organisation, and T for technology. (This was an improvement on Paul and Anne Ehrlich's I = PAT: population, affluence, and technology.)

The United Kingdom, with a population of 66.6 million and an area of 209,331 square kilometres, could fit comfortably into Victoria (237,657 square kilometres, with a population of 6.4 million). To a visitor, Britain seems to remain a green and pleasant land, achieving far more successful spatial management than Australia. Counties such as Hampshire, Wiltshire, Oxfordshire, Worcestershire, Derbyshire, and Staffordshire still look like Constable landscapes. And yet it is likely that Britain's population density—and the country's proximity to Europe— in addition to its shocking delays in adopting tough precautionary measures, have been major factors in the country's high death rate from COVID-19.

Japan's 127 million people live in less than half the area of New South Wales. Belgium (11.5 million people) could fit into the federal electorate of Gippsland. The Netherlands (17.2 million) has only one-third of the area of New York State (19.4 million).

If population—Australia's or any other nation's—is to rise,

then something has to be given up. Our report recognised that each increase in population puts additional strain on vital resources (such as water, soil, environment, open space, and transport), so that every million that the population increases by would, and should, have an impact on the way we live—on our use of cars, fuel and water, and space, and even on our diet. If we were prepared to adopt resource parsimony, we could accommodate more people. If not, we couldn't. Meanwhile, arid countries also face an existential threat, from desertification and permanent drought.

Climate change, displacement of people, and the refugee crisis

In his Leslie Stephen Lecture, 'Liberalism, populism and the fate of the world', delivered at Cambridge in 2018, Simon Schama, familiar for his books and television programs, makes a powerful case for linking climate change and the international refugee crisis, two major destabilising factors in contemporary politics:

> Look at our world now—with the eyes of a long-lens historian— and it's a truism that we are beset with three immense problems: the slow degradation if not slow death of our planetary habitat; the drastic inequalities of subsistence between the developed and undeveloped world; and not least the great gulf between those who wish only to live with people who look, sound, dress, speak and pray (if they do at all) like themselves, and those who are happy to live alongside people who aren't. In fact, all these conditions are functionally linked. Climate change has destroyed entire ecosystems, radicalising its casualties—long years of drought did that in the upper Jordan basin sending migrants into Syrian cities that couldn't find work for them and ending as recruits for both sides in the most terrible of contemporary civil

wars; famines in sub-tropical Africa have created the multitudes of desperate people setting off for Libya and calamity; the same is true of Central America and the tragic odyssey north. And eventually—this I suppose is the good or at least non apocalyptic news—all the walls, fences, and moats in the world will not be capable of dealing with the deep roots of the disaster and the vast waves of transhumance it has generated.[1]

Historically, 'transhumance' was defined as the seasonal movement of livestock seeking better pastures, but Schama's use, referring to forced human migration following the destruction of habitat (as shown in my following diagram), is compelling.

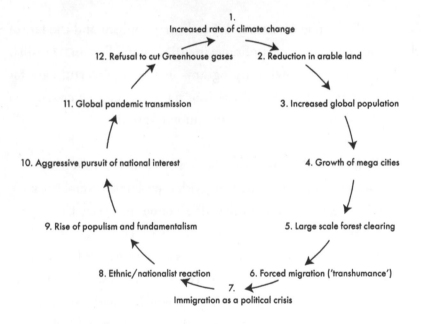

1. Increased rate of climate change
12. Refusal to cut Greenhouse gases
2. Reduction in arable land
11. Global pandemic transmission
3. Increased global population
10. Aggressive pursuit of national interest
4. Growth of mega cities
9. Rise of populism and fundamentalism
5. Large scale forest clearing
8. Ethnic/nationalist reaction
6. Forced migration ('transhumance')
7. Immigration as a political crisis

Resource-rich nations such as Australia are on the edge of a moral abyss. If our exports of coal and gas lead to China and India emitting greenhouse gases at even one-third per capita of Australia's rate, it will make us richer in the short term, but create a planetary disaster for our descendents in the medium to long term.

And yet Australia and the United States expose themselves as monumentally hypocritical if they say to China, India, Brazil, and Africa: *As rich countries, we will continue to consume natural resources without limit, but you must not follow our example.*

Increased longevity and dramatic falls in infant mortality are wonderful achievements. The threat to the Earth's atmosphere, water, soil, and biota is not caused by population in isolation—the numbers of people—but by the rate of per capita resource use, compounded by the creation of mega cities with their huge waste-disposal problems.

Cities are the major contributors to climate change. They are also the great vectors of pandemics, where social distancing is impossible except for very short periods. Climate change reduces the amount of arable land with secure water supplies, and the demand for food may far exceed our capacity to provide it.

Zoonoses are infectious diseases caused by a virus or other pathogens that jump from animals to humans, and COVID-19 is a zoonosis. Their devastating impact is closely related to climate change. The expansion of the human population, combined with the pressure to make more land available for food production, destroys the habitat of many animals and brings them physically closer to us, as a resource, with all their viruses attached, as celebrated English primatologist and anthropologist Jane Goodall has argued.

Bats are known to be vectors of corona viruses generally, while in parts of Asia rare animals are thought to have curative properties. Too many people believe that everything in nature can be consumed, without much thought for the consequences.

Climate change contributes to the refugee crisis, and to sources of conflict in the Middle East and Africa. Domestically, racial tensions that can be constrained during eras of stability and rising prosperity explode in times of crisis.

Europe, Canada, and New Zealand understand the connection between climate change and the threat to critical resources, and China does, too, up to a point. The United States, India, Brazil, and Australia do not. In Australia's case, the state governments are taking action on climate change: the Commonwealth refuses to act because, as have argued, of the corrupting power of vested interests.

There has been a retreat from globalism and liberal democracy at precisely the time when the planet faces existential problems that can only be solved/tackled by increasing international co-operation, raising levels of awareness, seeking out the best information/ evidence, and sharing it and acting on it. Retreating behind national boundaries, either physical or psychological/cultural/cognitive, and shouting patriotic slogans, is a recipe for disaster.

The World Health Organization has described climate change as 'potentially the greatest health threat of the 21st century',[2] and Oxfam has calculated that people are three times more likely to be displaced from their homes by cyclones, floods, or fires than by wars.

The World Bank has estimated that by 2050 more than 140 million people will have been displaced from sub-Saharan Africa, South Asia, and Latin America. They will be climate refugees. Even now, millions have been displaced by the reduction in arable land in North Africa, with thousands drowning in the Mediterranean as they have sought refuge.

Climate refugees were estimated to total 7 million people in 2019. The United Nations estimates that the number of refugees in 2050 will be far higher—at least 200 million, up to a billion. Their very existence and their expectations for life and security have created the greatest challenge to liberal democracy in the West— even greater than climate change itself, with which it is inextricably linked. One million refugees in central America and Syria have destabilised the politics of North America and the Middle East.

Their misery has failed to generate compassion and understanding. They are not protected by the 1951 Refugee Convention (The Convention Relating to the Status of Refugees), which defines a refugee as

> any person who owing to a well-founded fear of being persecuted for reasons of race, religion, nationality, membership of a particular social group or political opinion, is outside the country of his/her nationality and is unable, or owing to such fear, is unwilling to avail himself/herself of the protection of that country.

In December 2018, at the UN-sponsored Intergovernmental Conference to Adopt the 'Global Compact for Safe, Orderly, and Regular Migration', held in Marrakech, Morocco, an attempt to broaden the refugee convention to include climate refugees was wrecked by the United States and Brazil. Australia abstained.

Pandemics

Two familiar quotations keep churning through my mind: 'One death is a tragedy, one million deaths is a statistic', attributed to Josef Stalin (1943); and 'What is urgent will always take precedence over what is important', Edward de Bono, *Future Positive* (1979).

The coronavirus (COVID-19) pandemic, as I argued in Chapter 12, suddenly became the dominant priority of world leaders and their peoples (often after a period of denial). The pandemic was revolutionary in its impact, and forced governments to act decisively, often in unexpected ways, overturning strong and long-held political ideologies.

The striking contrast in the way governments have addressed the COVID-19 pandemic, which is personal, immediate, and local in its implications, and climate change, which potentially has catastrophic planetary consequences in the medium term, deserves our attention—and our action—during every waking moment.

In the unfolding coronavirus crisis, emergency financial support, provided by all governments at an unprecedented level was, at least initially, in response to need. The changes in direction provided some hope that, instead of dithering, denying, and delaying, leaders could rise to challenges and grow into their jobs, instead of retreating into a torpor, or playing an increasingly grotesque reality-television role, looking only to the next election and the applause of the stage-managed crowd.

In Australia, the most encouraging sign was that the Commonwealth government sought expert advice, applying scientific methods and rigorous testing, and engaged in serious consultation with states and territories, the opposition, the Business Council of Australia, and the Australian Council of Trade Unions, rather than turning only to lobbyists, pressure groups, vested interests, major party donors, shock jocks, and media barons, and reacting opportunistically.

Could the government do the same with climate change? Was the prime minister, Scott Morrison, capable of a Nixon-in-China moment that would shock the National Party and Pauline Hanson, but astound the world? Generally, Australia avoided the excessive politicisation of COVID-19, including eschewing xenophobia over the pandemic as a 'foreign' attack.

In 2020, the COVID-19 pandemic led, in a matter of months, to fundamental changes in how we live, work, feed, clothe, educate, transport, and entertain ourselves; transformed the role of government intervention; stretched medical and hospital services to the limit; inflicted lasting damage on the aviation industry and tourism; barred attendance at sporting events, funerals, and weddings; crippled universities and the peforming arts; devastated part-time and insecure jobs; widened inequality; and ran some risk of inflaming anti-migrant sentiment. It also profoundly disrupted the world economy.

Given the bloody history of the twentieth century, I assumed that nations might have become desensitised to large numbers of deaths. In World War I, about 22 million people were killed (most of them combatants); the Spanish flu killed about 50 million; in World War II, about 70 million (many non-combatants) were killed; in the Vietnam War, about 3.5 million; HIV-AIDS caused the deaths of about 32 million; and the Asian flu and the Hong Kong flu killed about 1 million each.

Politics was a direct cause of famines around the world last century: in China (1959–61, when 20 million people died); in the Congo (1998–2000: 2.7 million); in Bengal (1942–43: 2.1 million); in Biafra (1968–70: 2 million); in Cambodia (1975–79: 1.5 million); in North Korea 1994–98: 1.5 million); and in the Sahel, the climate zone between the Sahara to the north and the Sudanian savannah to the south (1968–72: 1 million).

Stalinist Russia was responsible for the deaths of many millions of Russian civilians and non-Russians. World War II was engineered by Hitler, and his fanatical anti-Semitism led to the Holocaust, in which 6 million Jews died.

But I was wrong about contemporary desensitisation. With COVID-19, the number of local deaths, and controversies about the correlation between the authorities' speed of response and low mortality rates, became central.

In the ANU's election surveys, voters who regarded global warming (ANU's choice of words) as a serious threat rose from 55 per cent in 2010 to 68 per cent in 2019: those who did not, fell from 45 to 33 per cent.

Climate change, unlike COVID-19, is not a new threat. There have been increasingly strong warnings for decades, but effective action is invariably postponed, although levels of government rhetoric remain high.

Its danger, including death, is potentially far greater but psychologically more remote, perhaps with decades before it becomes irreversible; its victims do not yet have a face or name attached to them; the science is complex, and some elements of it are disputed; there are powerful denialist lobbies, expressing unremitting hostility, year after year, especially in the mass media; and no leaders have emerged on this issue, while several have been destroyed by it. And the science is being ignored or discounted: *Leave it to the market. There's no role for government.*

In addition, personal conduct will not change outcomes—national, institutional, and corporate action will be required to save the planet.

But in a worst-case scenario, if global warming of 2 degrees or more occurs by 2050, famine on an unprecedented scale has the potential to kill hundreds of millions. But not in Australia. Unlike COVID-19, it will not be an urban problem—it will affect all the systems that enable our biota to operate. Pandemics will become more frequent and more deadly.

The planet is also threatened by the disposal-of-waste problem, but government responses have been tokenistic. About 20 per cent of food bought in Australia—and other rich countries—finishes in landfill. There has been some attempt to limit the use of plastic, but its damage in pelagic regions and to the life of fish, whales, and birds, compounded by wastewater outflows from sewage, industry (mostly mining), and chemicals used in agriculture may be irreversible.

As a 2018 BBC documentary has demonstrated, seabirds on Lord Howe Island, one of the most remote places on earth, are starving to death because their stomachs are so full of plastic that there is no room for food.

But although such issues are in the blindingly obvious category, they are not on the political agenda because they are covered by a political vow of silence by all major parties—challenging the use of plastics and chemical herbicides is seen as a threat to growth and to jobs, and offends powerful lobbyists.

There is common ground on diagnosis of the problems: the difficult part is providing solutions. The only way in which our political paralysis can be ended and effective action taken to save the planet is by active, forceful political engagement by well-informed citizens, changing the political culture and challenging the existing parties.

They must engage, engage, engage.

CHAPTER FOURTEEN

What Is to Be Done: political engagement and climate change

To abandon facts is to abandon freedom. If nothing is true, then no one can criticize power, because there is no basis upon which to do so. If nothing is true, then all is spectacle. The biggest wallet pays for the most blinding lights.

– TIMOTHY SNYDER, *ON TYRANNY* (2017)

[W]hen people ask about the meaning of life, they expect to be told a story. *Homo sapiens* is a storytelling animal, that thinks in stories rather than in numbers or graphs, and believes that the universe itself works like a story, replete with heroes and villains, conflicts and resolutions, climaxes and happy endings.

– YUVAL NOAH HARARI, *21 LESSONS FOR THE 21ST CENTURY* (2018)

The facts? The facts? I have been elected to change the facts.

– MARGARET THATCHER (1979)

As I have argued, climate change and other major problems will have been caused by us and our contemporaries. But we will not see the outcomes of our greed and thoughtlessness: those who will suffer the worst will not have caused the problems.

When Scott Morrison declared a belief in miracles, and happy endings, and when Donald Trump reassured Americans that climate change is a hoax and COVID-19 is '99 per cent totally harmless' and that one day it would just disappear, these were part of a narrative that people wanted to hear—a comforting bedtime story.

Since both major parties in Australia and the United States have fallen out of the habit of arguing rigorously and engaging in serious debate, it is understandable that people look for the reassuring narrative.

Trump is a fabulist whose narrative encouraged a large cohort of Americans but was mysterious or infuriating to everybody else. But Trump was not talking to Europeans or Australians. Why bother about the rest of the world? They don't vote in American elections, although in reality they have become non-voting constituents.

In contemporary society, developing a narrative has greater psychological carrying power than relaying the facts provided by experts. Subjective factors outweigh objective ones, notwithstanding our higher levels of formal education. As the quotations above suggest, there may be little common ground.

We should embrace the challenge of complexity, stretch our intellectual resources, learn how to process information, and preference evidence over feeling.

Information is available in abundance, but its validity must be subject to rigorous testing. We should never rely on sources simply because we are comfortable with them. The World Wide Web is a false friend when it delivers material to us that is tailored to exploit our fears and enthusiasms.

After years of agonising, I have concluded (but I am hardly Robinson Crusoe; other writers have worked their way to similar conclusions) that these are the priorities for our time if we are to survive the next half century without irreversible damage to the biosphere and our social and political institutions:

1. Take strong action on climate change, transition to a post-carbon economy, recognise the vital role of science, collecting evidence and applying scientific method in decision-making, acknowledging that coal is the biggest single source of greenhouse gases, that the environment is not the enemy and that Australia has the capacity to be a world leader in economically complex industries.

2. Challenge major parties to adopt open democratic practices, come clean on funding, expose the role of lobbyists, and restore trust in public institutions. If the major parties fail to respond, citizens will have to create alternatives.

3. Reject 'the Nixon strategy' of winning elections by promoting division, cultivating 'the base', persuading economic victims to blame those below them, exploiting the condescension factor, stoking resentment of expert opinion, and adopting 'winner take all' ('the people have spoken') practices to discount dissenting opinions.

4. Commit to strengthening liberal democracy and recognise the threat posed by the rise of populist authoritarian leaders, Xi, Putin, Trump, Modi, Bolsonaro, Erdoğan, Orbán, Duterte, López Obrador.

5. Protect the right to be informed as a central tenet of democracy: preserving the ABC and its investigative reporting, recognising the importance of a free press, not subject to commercial pressures or government bullying; strengthening the public service's capacity to give 'frank and fearless' advice; and providing adequate funding for the CSIRO, the Bureau of Meteorology, and university research. Resist secrecy as the instinctive fallback position by government to avoid embarrassment.

6. Recognise that inequality is not just a social Darwinian by-product of the economic system ('survival of the fittest') but a political artefact: not an accident, but built into decisions

about taxation, education, and health. Thomas Piketty is right (Andrew Leigh, too).

7. Recognise that the goal of education must be to enable people to fulfil human potential for the whole of life, not just to train pupils to be consumers and producers for the contemporary economy. The syllabus should include some political science, philosophy, the humanities, the arts, and exposure to comparative religion, encouraging curiosity, imagination, creativity, a grasp of historical perspective, aesthetics, and close exposure to nature.

8. Reject all forms of racism and adopt rational and humane responses to the refugee/asylum seeker crisis. Give high priority to the 'Uluru Statement from the Heart', and take stronger action on the Closing the Gap strategy.

Political engagement: the 0.2 per cent problem

> The only thing necessary for the triumph of evil is for good men to do nothing.
> – EDMUND BURKE (1770)

A robust democracy depends on high levels of citizen engagement, and this demands an investment of time, energy, commitment, knowledge, judgement, and balance. Short of armed revolt, which I would not recommend, it is the only way that our system can be reformed to restore the concept of public office as a public trust, and to preference the public interest over vested interests. Are we up to the challenge? And are our schools and universities doing enough to explain how democracy works, and how institutions interact with our lives?

Trust has been declining in the democratic system, and in public institutions—government, parliament, the political parties, the media, churches, schools, police, banks, insurance companies—and even in sport. It is understandable that Australians might be giving up on democracy, since our major political parties abandoned it in their internal processes years ago. However, COVID-19 has checked this decline, at least briefly.

The current cohort of Australians is the most highly qualified in our history, with 6.9 million graduates, and before COVID-19 more than 1 million were undertaking tertiary studies. Among this vast cohort are many with outstanding, internationally recognised, expertise in many disciplines—including medicine, law, agriculture, metallurgy, astronomy, climate science, oceanography, the humanities, the performing arts, engineering, and philosophy. And yet universities have become trading corporations (they had no alternative), many graduates are working their way through existing market structures, and superspecialisation has led to a high degree of fragmentation, with a primary emphasis on the individual's career, and wealth often seen as a goal in itself.

In addition, the overwhelming majority of Australians are disenchanted, disillusioned, or repulsed by our politics, and take no active part in reforming our political systems and changing our priorities.

Disturbingly, although we have the highest proportion of graduates in our history (six times more, *pro rata*, than in 1982), we have opted for banal timidity and political disengagement. We have discovered that education is not necessarily liberating. For some, it provides access to economic power without enhancing insight.

Of 15 million Australian voters, barely 30,000 have even a nominal involvement in political parties—an engagement of just 0.2 per cent. The parties are small, closed, secretive, and oligarchic, and they prefer it that way.

If, instead of engaging in handwringing and voting with pegs on their noses, 10 per cent of voters joined the political party that they generally voted for and played an active role in policy formulation, they could transform Australian politics very speedily. In practice, even 5 per cent (750,000) might be enough to do so.

This is a modest figure, proportionately, compared to the period after World War II when party memberships were high. But it would transform politics beyond recognition, change party structures, and lead to more courageous policies. The problem for all major parties is that their traditional bases are contracting, so they come to rely on zealots and lobbyists, selling their integrity to the highest bidder.

The hegemonic parties discourage a large membership, contrary to what might be expected, because the people who currently run/control/own the parties are unwilling to open up internal debate on policy, and they do not want to lose control.

And engagement needs to be direct, personal, and face-to-face to be effective. As Oscar Wilde reputedly observed, 'The trouble with socialism is that it takes up too many evenings.'

GetUp! spent $4 million in the 2019 election, targeting seven Coalition MPs, but had only one success: the defeat of Tony Abbott by Independent Zali Steggall, who would probably have won anyway. It is comparatively easy to secure thousands of signatures online, but effective activism requires more time commitment than just depressing the 'Send' key. GetUp! claims to have about 1 million subscribers, and the policies it promotes are admirable, but its operations may be too diffuse to be effective. GetUp! should think about Archimedes and his lever.

Just four Australians—Rupert Murdoch, Gina Rinehart, Andrew 'Twiggy' Forrest, and Clive Palmer—between them (or even individually) have far more political influence than GetUp!'s 1 million supporters.

In the United States, a central element in defining political freedom is the right *not* to be involved and *not* to vote. Having very low turnouts can be a matter for celebration. Donald Trump, in a rare moment of insight, observed that if there was a high voter turnout no Republican would ever be elected—so voter suppression (and gerrymandering) is a very important strategy in the 'red' states.

Australia is different. More than 90 per cent of us turn up to vote—but that's as far as our engagement extends. Lining up to vote is compulsory (something the Americans regard as inexplicable) but we take it for granted. In New Zealand, where voting is not compulsory, there was a 79 per cent turnout in 2017 (with younger voters not participating).

Citizens queue up at polling stations (or mail in their votes) every three years (and sometimes less) in Commonwealth elections, and every four years in every state and territory except Queensland (three years). They pay their taxes (except where they receive professional advice on how to avoid them), but keep well out of the political process.

On the political menu, consumers (that is, voters) basically have the alternative of McDonald's or KFC. They might prefer something other than a Big Mac or fried chicken, but they must turn up and choose, whether they like it or not.

Few Australian citizens recognise that our federal parliament meets for shorter periods than any other national legislature in the Western world (see p 239)—and that state parliaments sit for even fewer. So the idea of open, sustained debate and rigorous questioning has become just a nostalgic memory. Executive government is increasingly irritated by any intervention by the courts to constrain its authority—and much activity has been specifically excluded from judicial scrutiny. There is no federal Independent Commission Against Corruption. Governments generally—and the government in particular—deeply resent investigative journalism by the ABC or what were formerly known as the Fairfax newspapers. The role of the public service has been diminished.

The concept of 'checks and balances' in our political system is crashing down. The rise and rise of the executive has been accompanied by a reduced role for parliament and significant policy areas being excluded from judicial review. Australia has become one of the world's most secretive democracies.

The principal role of the House of Representatives (like the Electoral College in the United States) is, in effect, to certify the election of the executive, and then retreat. Backbench MPs don't make policy these days—although they can break them—and they don't debate serious issues. But they do have time to plot, the only area where they can have significant influence. Politics has become almost entirely personal, constructed around fiefdoms and alliances, and the manipulation of hatreds.

Voters participated in the change of Australian prime ministers in 2007 (from Howard to Rudd) and 2013 (Rudd to Abbott), but not at all in 2010 (Rudd to Gillard), 2013 (Gillard to Rudd), 2015 (Abbott to Turnbull), and 2018 (Turnbull to Morrison).

According to the Australian National University's *Trends in Australian Political Opinion 1987–2019*, Abbott's removal in 2015 in a slow-motion coup was approved by a slender majority (51 per cent), but there was strong opposition to the other changes: 74 per cent in 2010; 58 per cent in 2013; and 74 per cent in 2018.

Overseas satirists describe Canberra as the coup capital of the Western world.

Voters are now spectators, not participants, in the political process, in which the real and the virtual have been inverted, as if a horror movie represents the reality, and the audience cannot change the outcome.

Party structures are oligarchic and secretive, and their members, in practice, comprise two categories: insiders (being small in numbers, but powerful) and outsiders (in larger numbers, but ageing and weak).

Factions, trade unions, industry groups, and substantial donors are major players. So are the media, especially News Corp, with by far the greatest spread of newspaper readership, and lobbyists—for example, the fossil-fuel lobby and its allies, such as the Institute of Public Affairs and the Centre for Independent Studies; the gambling industry; property developers; and private schools.

There are also public-interest lobbyists, including GetUp!, the Australian Conservation Foundation, the Australian Medical Association; think tanks (such as the Grattan Institute, the Lowy Institute, and the Australia Institute); the five learned academies; the universities; the churches; sporting groups; and public-interest media—such as the ABC, SBS, Crikey, Schwartz Media, and the former Fairfax newspapers. Sporting groups such as the Australian Football League and the National Rugby League have large memberships and considerable lobbying power.

Climate change is the most egregious example of an important policy having been hijacked by vested interests, intent on promoting the expansion of the fossil-fuel industry whatever the environmental costs. And yet, although there are millions of Australians with lived experience of climate change from direct observation—farmers, gardeners, vignerons, photographers, aviators, birdwatchers, bushwalkers, anglers, skiers, and beekeepers, as mentioned in Chapter 7—their expertise has not been called upon.

There is no serious attempt at engagement by government. The lobbyists win by default.

The role of the lobbyist is like the hidden part of an iceberg. Vested interests are very sophisticated in going after what they want, while defenders of the public interest are often guileless and naïve.

Australia handled the first wave of the COVID-19 crisis exceptionally well but, inevitably, all the major decisions were made by executive government, relying on expert advice. There was nothing particularly 'democratic' about it. Parliaments, both federal and state, were only peripherally involved, except to wave through massive appropriations to respond to the economic consequences of lockdown. This was a further striking demonstration of a breakdown in the separation of powers.

The government and the opposition worked together to tackle the coronavirus emergency, and this co-operative relationship was welcome. But it proved to be short-lived. Scott Morrison was looking for a sudden 'snap-back' (his words) and a return to the 'old normal', essentially *Thanks and goodbye, experts!*, and *Welcome back, lobbyists!* Parliament itself played only a vestigial role in the scope of the emergency measures, however necessary they were, not even asking critical questions about expenditure—which were at a level never seen before.

By early August, Victoria, with 27 per cent of Australia's population, had conducted 37 per cent of the COVID-19 tests, but had 69 per cent of the cases and 71 per cent of the deaths. In the first wave, early restrictions in the state had kept the infection rate low; but in the second wave, case numbers grew dramatically, and Victoria declared itself 'a state of disaster.' This is ironic because Melbourne is one of the world's great centres for medical research.

Victoria was isolated by strict border closures, and the co-operative mood in dealing with the first wave broke down, with 'compliance fatigue', partisan attacks, and/or damaging briefings against Dan Andrews and his government by federal ministers. The areas of concern were bungled hotel quarantining for returning travellers, infections and deaths in privatised aged-care homes (strictly, a Commonwealth responsibility), problems in abattoirs, some schools, and ethnic communities, and a lack of follow-up after testing. 'Aged care' became an oxymoron. Insecure workers needed income even after infection, and failed to self-isolate.

Unprecedented powers were given to the Victoria Police, which was not trained to deal with disease, not subject to judicial review, and could invade houses and impose fines, and the police minister was given authority to suspend legislation temporarily. However, unlike 1793, a Committee of Public Safety was not set up.

At a time when many countries were redefining and limiting police power, Australia generally, and Victoria in particular, was pushing in the other direction.

New Zealand had 102 days without new infections and hoped to avoid a second wave, but Jacinda Ardern's mantra 'Be kind' had no appeal to Australian leaders, confirming an authoritarian and populist streak in our history.

There has been a striking crossover in political allegiances in recent decades in the United States, the United Kingdom, France, Canada, Australia, and New Zealand, in which educational levels, rather than income levels, are more likely to indicate party preferences. The phenomenon of the Audi- or Tesla-driving professional who votes for the ALP is rising, while tradies with a Toyota ute have been on a long, but steady, transition towards the Coalition, especially if they have invested in negatively geared property. Being part of an elite is now a term of political abuse—an accusation of remoteness, condescension, and not being (in Australia anyway) 'fair dinkum'.

Rejecting concepts of hierarchies of learning or experience— what could be more democratic than that?

Populism has become a significant factor in Australian politics, and the Coalition depends on preferences from a hard-right voting group to win in tight elections.

In the Victorian state election of November 2018, the ALP secured 1,506,410 primary votes (42.9 per cent), and in the federal election of May 2019, 1,386,078 (36.9 per cent). Even if we take the lower, 2019, voting figure, and accept the probably inflated 16,000 membership claimed for the Victorian ALP, this suggests that the party can only attract 1.15 per cent of its supporters to join, while a mere 98.85 per cent prefer to opt out. And this in the ALP's most consistently successful branch!

'There's 30,000 of them and only 15 million of us': looking for alternatives

If 5 per cent of Australian voters, the estimated 750,000 mentioned above, joined—or even attempted to join—existing parties, they would blow open the entrance to some dark caves. But if the hegemonic parties rejected them, they would then have the option of retreating, or forming a new political force—perhaps a Courage Party.

Party apparatchiks are preoccupied with preserving the vehicle, ensuring its electoral support, and less interested in the destination, especially if it is over the hills and far away.

I joined the Australian Labor Party in 1950, just 70 years ago, and would now classify myself as an anxious life member. I owe the ALP a great deal, particularly for my 26 years as a member of parliament, both state and federal, and seven years as a minister. But during a period of dramatic global change with profound implications for Australia and liberal democracy generally, all Australian political parties have demonstrated their inadequacy.

Apart from a commitment to 'fairness'—rather a vague concept—the ALP has become very risk-averse, retreating to its historic base, failing to build on the radical innovations driven by Chifley, Whitlam, Hawke, Keating, Rudd, and Gillard. However, some Labor state premiers have been effective, even taking personal responsibility on some issues, which is a rarity.

It is hard to think of an issue that Labor would not modify or abandon under pressure.

If there was a *Truth in Politics Act*, existing parties could be required to adopt new, more accurate names, for example: The Self-Interest Party; The Coal Party; The Tepid Party; The Pure Party; or The Me Party.

Courage attracts; lack of courage repels. I would like to see a political party that was committed to both economic and non-economic objectives, with a strong emphasis on moral values, broadly defined, and truth telling. Beyond the obvious imperative of taking strong action on climate change and moving to a post-carbon economy, a new party should commit to:

- renovating or replacing the Australian Constitution (1901), building on the model adopted in Queen Victoria's time, to include references to cabinet government, democratic practice, modified by 120 years of history, recognising Indigenous rights and an Australian head of state;
- adopting a Bill of Rights;
- creating an Independent Commission Against Corruption, reporting directly to the Parliament;
- open government, judicial review, and rejecting the imposition of secrecy to save governments from embarrassment (citing 'national security' as the excuse for suppression);
- appropriate funding for trusted Australian institutions, including the ABC, the CSIRO, the Bureau of Meteorology, and the Australian Bureau of Statistics;
- investing in 'soft diplomacy' in the south-west Pacific, including broadcasting to the region, and the provision of medical, scientific, and infrastructure support and disaster relief;
- recognising the value of the humanities and the arts, and providing adequate funding for them;
- giving high priority to science as a key to understanding how the world works, and adopting the scientific method in decision-making, including economic goals
- re-establishing a moral basis for progressive taxation;
- acknowledging the increase in social inequality and the role of taxation and school funding as contributing factors;
- providing Commonwealth funding for schools on the basis of need, ending favouritism towards particular sectors, and ending

the political weaponisation of parental anxiety about school choice;

- revising school syllabuses to include some understanding of democracy, ethics, philosophy, tribalism, patriarchy, misogyny, and colonial history;
- rethinking foreign, defence, and trade policies;
- being open-minded and sceptical about involvement in foreign wars (Vietnam, Afghanistan, Iraq) and super-charged spending on military hardware (especially submarines) without rational analysis and full parliamentary debate;
- taking more creative approaches to resolving tensions arising from multiculturalism;
- rejecting the notion of education, health, and the arts as industries;
- reviving the concept of 'the public interest';
- adopting compassionate, rational, and statistically based policies towards refugees/asylum seekers arriving by sea; and
- establishing openness in the party's own procedures.

I would like the ALP to be that party. Parties sometimes go through the motions of attempting to reform internal processes—but a paint job will not suffice.

Of course, setting up an effective alternative political force would acquire enormous sacrifice and cost, with dedicated citizens putting global and national interests ahead of their own careers.

Australia's COVID-19 success demonstrated that the system is capable of working well where there is mutual co-operation and trust. But early indications are that 'post-COVID-19' will mean a return to the status quo.

Leaving aside the vehicle (political parties, however constituted) and concentrating on the destination, these are the national priorities that must be adopted both for Australia and globally—and in the medium to long term, they are inseparable.

Take strong action on climate change and transition to a post-carbon
economy

Professor Ross Garnaut is Australia's pre-eminent expert on the economics and social impact of climate change, and has had a greater impact on policy formulation in the post-war era than any Australian economist or social scientist since 'Nugget' Coombs. In *Superpower* (2019), Garnaut identifies 'four great opportunities for Australian industrial leadership in the post-carbon world economy: globally competitive renewable power; use of competitive electricity and hydrogen for local processing of a high proportion of Australian mineral production; an abundance of biomass for the chemical manufacturing industries; and low-cost biological and geological sequestration of carbon wastes'.[1]

He adopts the conclusions of the CSIRO's *Climate Change: science and solutions for Australia* (2011) that Australia has great potential to store more carbon in soils, pastures, woodland forests, and biodiverse plantations, which could mitigate about 20 per cent of emissions.

He argues that adopting his proposals would lead to a 50 per cent reduction of emissions (on 2005 levels) by 2030, and would be a 'credible launching position for zero emissions before 2050'.

At first, his advocacy for more aluminium and iron smelting in Australia, both dependent on huge electricity inputs, seems counter-intuitive, but he estimates that 'converting one-quarter of Australian iron oxide and half of aluminium oxide exports to metal would add more value and jobs than current coal and gas combined'.

'Direct reduction' is a technology in which the conversion of iron oxide into metal uses hydrogen (made from renewable energy) as a 'reductant' (a substance that brings about a reduction in another substance while it is oxidised itself). After hydrogen pulls oxygen atoms out of iron oxide, it leaves water vapour, but not carbon dioxide, as the waste, essentially creating a zero-emissions path.

With aluminium, replacing coal-fired power with renewables in the electrolysis process would lead to zero emissions. Fossil-carbon-based anodes and cathodes, which gradually oxidise in the electrolysis process, would need to be replaced by renewable carbon or other materials. Both Alcoa and Rio Tinto are developing this technology in North America.

'Arid and semi-arid rangelands make up about 70 per of Australia's landmass, or around 5.5. million square kilometres', he writes. This is no good for farming, but ideal for growing billions of trees with a high capacity to absorb carbon dioxide—especially mallee eucalypts, or mulga (*Acacia aneura*) in Queensland and New South Wales. This would be, by far, the most cost-effective form of carbon capture and storage. 'Landowners will think hard about the parts of their properties that would have more value as carbon sinks than carrying sheep.'

Biomass could displace coal as a fuel in some areas, and Australia has exceptional potential here. There is, however, a significant difference of opinion about its use as a fuel. Critics point out that biomass is less carbon-dense than coal, so greater amounts would need to be burned to produce an equivalent amount of energy, and large storage areas might be needed because of its high volume. Advocates argue that biomass is carbon-neutral, only releasing the carbon dioxide absorbed during its lifetime. Biomass would normally be consumed close to where it was grown, reducing transport emissions.

Professor Garnaut sees the role of biomass as a replacement for fossil carbon in chemical-industrial processes as potentially more significant than in electricity generation, where the costs of solar and wind are falling. The soil has enormous potential for carbon absorption, probably more than trees, and, as he writes, 'the application of charcoal or char to magnify increase in carbon has soil productivity as well as sequestration benefits'.

Much more needs to be done to promote reductions in waste, and increases in energy efficiency and recycling. Garnaut argues that Australia has the skills to become a global leader in these areas—the bigger question being whether we have the will and the leadership to commit to the future, rather than to keep shoring up the past.

Garnaut emphasises the importance of setting a carbon price, but only if there is political consensus to do so, to provide certainty and security for the market.

He sees the 'harvesting' and containment of human and animal wastes as a major renewable source of energy, in addition to Australia's steady increase in the use of solar panels and wind farms.

Nuclear power has one significant advantage: it does not produce greenhouse gases, except in the construction phase. But quite apart from safety concerns in the post-Chernobyl era, and the dangers associated with transporting nuclear fuel from the source to the plant, economics make it non-competitive (even though Australia has a lot of uranium to sell). The costs of building a nuclear plant would be enormous, as would decommissioning it at the end of its life and burying all the materials involved.

Australia's chief scientist, Dr Alan Finkel, in his 'Hydrogen for Australia's Future' briefing paper (2019) proposes that, 'When it comes to capturing and exporting solar and wind electricity by first splitting water into hydrogen and oxygen, clean hydrogen and its derivatives have no equal. Energy importing countries are hungry for hydrogen as part of their emissions reduction agenda, and Australia has the potential to supply much of their needs.' Using solar energy, hydropower, and other renewable sources for electrolysis would produce a fuel with high export potential and no greenhouse-gas emissions.

Research originating in Australia with Meat and Livestock Australia, James Cook University, and the CSIRO gives encouraging indications that the emission of methane by cattle—a major source of greenhouse-gas production—can be reduced by more than

50 per cent by feeding them a 1 per cent dietary supplement of the common Australian red seaweed, *Asparagopsis taxiformis* or *armata*, at a particular stage of its development, as a red-coloured sphorophyte. Metabolites (small molecules) in the seaweed disrupt enzymes that produce methane in the rumen of cattle.

This research is also being developed at the University of California at Davis and San Diego, Stanford University, and the Scripps Research Institute at La Jolla, California.

The *Asparagopsis* sphorophyte will have to be extracted from the oceans, then grown in commercial quantities. And to be effective, cattle must have it in their feed every day. Its implications for beef cattle and the dairy industry are enormous, provided that the economics, including the cost of production and delivery, can be reduced with scale.

CSIRO research also demonstrates that feeding cattle two tropical legume species, *Leucaena* and *Desmanthus*, can both reduce methane and increase animal growth.

Soil Carbon Co., in Orange, New South Wales, is developing a program to capture carbon dioxide from the atmosphere on a gigatonne scale, using microbial fungi and bacteria, and to return it to the soil, which would improve soil health and allow farmers to trade carbon credits.

This is carbon sequestration and storage with only a tiny fraction of the huge energy inputs involved in the engineering model, and the added advantage of benefiting agricultural production.

It is, they claim, 'an elegant solution for two of our greatest challenges: the decrease of fertility and resilience in the world's agricultural soils, and climate change induced by the increase of carbon dioxide in the atmosphere'.

These are all encouraging signs.

Australia must lead by example, ranking as it does as no. 1 in the world for per capita emissions of greenhouse gases. The COVID-19 lockdown will have reduced urban emissions sharply, but the 2019–20 bushfires will have increased them.

We should use our expertise and newfound moral authority after our relative success with COVID-19 to act in concert with other nations to put pressure on countries such the United States, China, and India to cut back their emissions.

Government must take a moral lead, acting courageously, with conviction, and not feel weak at the knees at the thought of challenging the United States. As explained in Chapter 7, we can either take the lead on a transition to a post-carbon society or wait until it is forced on us. But our leaders have to become advocates. And they must be seen to mean what they say.

Remake the export economy: on to the 2020s

Australia is a highly sophisticated urban society, but it is dependent on high-volume exports of raw materials—usually the characteristic of third-world or developing economies. And this is not an accident. It is planned that way. Governments encourage it, and investors give their highest priority to mining and property.

Australia's spectacular failure to develop elaborately transformed manufactures has inevitably led to an over-reliance on fossil fuels for our export income, adding to the risk of irreversible climate change. Of our ten major export earners, all but two are raw materials or concentrates (iron, coal, natural gas, gold, aluminium, beef, petroleum, and copper). The only exceptions—foreign tourists and overseas students—are both under long-term threat due to COVID-19.

The criticism that Donald Horne made in *The Lucky Country* in 1964 that we think of Australia as a quarry and a farm is still valid: Where would we be if we only had the resources of Israel, Sweden, or Finland?

The National Party is still fighting to keep it that way. The Liberals are divided, and Labor is unable to make its mind up.

Harvard University's Center for International Development publishes an authoritative *Atlas of Economic Complexity*, based on the premise that 'Economic development requires the accumulation of productive knowledge and its use in both more and more complex industries ... Countries improve their Economic Complexity Index (ECI) by increasing the number and complexity of the products they successfully export.'

Because Australia has chosen to be primarily an exporter of high-volume raw materials and refuses to seize the opportunities arising from transitioning to a post-carbon economy, which would

ECI rankings	1995	2018
Australia	55	87
Japan	1	1
Switzerland	4	2
South Korea	21	3
Germany	3	4
Singapore	20	5
Sweden	3	8
United States	9	11
Finland	5	12
United Kingdom	7	13
Italy	10	14
France	8	16
China	46	18
Israel	19	20
Canada	22	39
India	60	42
Brazil	25	49
Vietnam	107	52
New Zealand	33	54
Indonesia	76	61

create highly sophisticated export industries, our ECI is falling. We have the skill to reverse this, but not the will.

The good news for Australia is that we are well ahead of Ethiopia and Papua-New Guinea.

Government support for STEM (Science, Technology, Engineering, and Mathematics) disciplines is welcome, but there is no overarching plan to develop new industries capable of driving the transition to a post-carbon economy. Indeed, all government rhetoric harks back to the 1980s.

Despite Boris Johnson's encouraging words, the export of Tim-Tams and Vegemite will not be enough.

Recognise that governments can play a useful role

One of the worst features of the Thatcher–Reagan jeremiad, taken up with characteristic paranoia by Trump, was the concept that government—and, especially, taxation—was the enemy of economic advancement and the creation of new businesses, and that the strong were being penalised to advantage the weak. In the neoliberal model, 'self-regulation'—a by-product of competition—was likely to be far more effective than government oversight, usually dubbed 'red tape.'

In Australia and elsewhere, self-regulation has proved to be a spectacular failure. The community has suffered heavy damage through the eclipse of values and integrity, as the Hayne Royal Commission into Misconduct in the Banking, Superannuation and Financial Services Industry demonstrated. Self-regulation has been directly responsible for a loss of trust: churches who might have been regarded as moral exemplars have been cruel, smug, and secretive, using all their networks against openness; banks, once seen as models of probity, have robbed their customers.

For decades, the concept of fairness, balance, and social equity has taken a low priority. It has been easy for corporations to avoid paying company tax. Inequality has been legitimised, and deprivation dismissed as the result of tough luck or personal inadequacy.

Government can play an important role by opening up areas for careful debate, collecting and providing testable information, setting goals with an ethical content and concern for the medium to long term, and making a commitment to preserving and enhancing the environment that we share. Accountability and transparency are central.

Mike Cannon-Brookes spearheaded the Beyond Zero Emissions (BZE) think tank (June 2020), which demonstrated that, by embracing a post-carbon economy, 1.8 million new jobs could be created in Australia in five years in energy, manufacturing, agriculture, construction, and transport.

Reject racism and rethink Australia's response to the refugee crisis
Australia had a long history of discrimination in immigration, dominated by the White Australia Policy, until the 1970s.

In July 1938, in France, at the Évian Conference on the refugee crisis, Australia took a very tough line against the boat people of the time—Jewish refugees from Hitler's Germany. Thomas White, the United Australia Party minister for trade and customs, said, '... undue privileges cannot be given to one particular class of non-British subjects without injustices to others ... as we have no real racial problem, we are not desirous of importing one.' He suggested that the Jews should remain in Germany and try to sort things out—not happy advice, as it turned out.

John Anderson, formerly deputy prime minister, National Party leader, and a practising Christian, was eerily channelling White on the ABC's *Q&A* program on 13 April 2020 when he encouraged refugees to go back home and try to fix things up politically.

Australia's approach to refugees is not just cruel, but arbitrary and selective. (It is also political pay-dirt for its advocates.)

Between 1975 and 1985, mostly under Malcolm Fraser's prime ministership, 132,000 Vietnamese arrived on boats, with bipartisan support, and were readily accepted into the Australian community.

The largest number of boat people in recent times—51,000—arrived in the period 2009–13: an exceptional spike over a 30-year period. But Kevin Rudd declared in 2013 that 'asylum seekers who come here by boat without a visa will never be settled in Australia', and offshore processing then became mandatory.

However, 72,000 asylum seekers arrived in Australia by air in 2016–19, under Abbott, Turnbull, and Morrison, according to an answer given by Peter Dutton to a Question on Notice. Those who arrive by air are comparatively well treated, and many drift into the community: the most punitive treatment is reserved for the dispossessed and desperate. Nobody, so far, has proposed that the RAAF be used to turn back commercial flights with potential refugees on board.

The United Nations High Commission for Refugees (UNHCR) reports:

> Between January 2009 and December 2018, Australia recognised or resettled 180,790 refugees. This represented 0.89 per cent of the 20.3 million refugees recognised globally over that period. Australia's total contribution for the decade is ranked 25th overall, 29th per capita, and 54th relative to national GDP.

Australians have never been exposed to a detailed and dispassionate presentation of the refugee issue, including the number of persons involved and the cost of institutionalised cruelty—$A3.3 billion per year, or $400,000 for each person in offshore detention—dealt with statistically and analytically. Nor is there comprehensive reporting of the numbers involved in other countries. This is in sharp contrast to how the COVID-19 emergency was handled.

Apart from the boat/air issue, there has been a failure to examine the development of Australia's authoritarian and extra-legal methods—imprisoning people administratively, instead of after a judicial process.

Our location puts us on the margin of the refugee diaspora, remote from the centre. Whereas Europe and the Middle East are facing millions of rtefugees each year, this large, rich, and (mostly) generous nation is potentially challenged by tens of thousands at most.

Practising Christians in our major parties have abandoned the parable of the Good Samaritan. Once we dehumanise others, either systematically or randomly, rendering them nameless, faceless, without any identity, history, or aspirations, we inflict psychic damage on ourselves. Treating it as normative, and finding excuses, is destructive.

Our leaders lack comprehension of what it is to be 'other': to have the empathy to understand or change roles with victims. We could begin by no longer referring to refugees as 'illegals'. Under the Refugee Convention, it is not illegal to seek asylum.

Worst of all is that institutionalised sadism has proved to be electorally popular and Oppositions are hesitant to attack it, so there is community acquiescence. The ANU's polling indicates that the Coalition's position on refugees and asylum seekers has been more popular than Labor's, and has been consistently so since 2001. That does not make it right. It simply means that a feeble argument strengthens the status quo.

Resist fundamentalism

Religious fundamentalism—both Christian and Islamic—seems to offer cheap grace, a superficial transaction promising lifelong, even post-life, guarantees, just like buying a commercial product such as life assurance. Questioning, individual judgement, or knowledge is not required, and may be actively discouraged.

Fundamentalism is not merely intellectually crippling; it is profoundly contemptuous of Jesus or Muhammad, whose teachings are far more profound, universal, stimulating, controversial, and compassionate than fundamentalists will concede. Fundamentalism offers a creed without history, without scholarship, without depth, and without context, and yet its phenomenal growth confirms that it meets community needs and anxieties far more than mainstream churches (or mosques).

After 2000, the most dynamic political force in the United States was a coalition of evangelical fundamentalists, the neocons (neo-conservatives), and corporate power, strongly supported by mass-media owners. This was not fascism in a European context, but there were disturbing ideological parallels, which Philip Roth took up in his novel *The Plot Against America* (2004).

And fundamentalism has become the dominant force in the Middle East, especially after Western interventions in Iraq and Afghanistan were recognised as tragic—and expensive—failures.

As Daniel Patrick Moynihan said, 'Everyone is entitled to his opinion, but not to his own facts.'

Reinforce the moral basis of progressive taxation

In 1927, Justice Oliver Wendell Holmes, Jr of the US Supreme Court, famously wrote, 'Taxes are what we pay for a civilized society.' This has been the case since the first cities were established around the Euphrates, more than 5,000 years ago,

The Commonwealth of Australia introduced its first income tax in 1915. Before that, most of its revenue was derived from tariffs, and the states collected income and land taxes. After 1942, the Commonwealth had a monopoly on income tax, but paid off the states with grants. During World War II, the top marginal rates were very high—more than 90 per cent.

Progressive taxes were the norm until the 1980s when the Thatcher–Reagan model of smaller government and devolution was adopted by other governments, and there were significant reductions in tax rates: from 60 per cent at the top to 45 per cent in 2010.

Australia, and comparable Western nations, are marked by three common factors: people are living far longer, often two or even three decades more than the previous generation; in the last decade of life, many people will require sophisticated and expensive medical intervention and/or institutional support; and with their increasing participation in tertiary and post-graduate education, young people will be entering the labour force later, in their early twenties.

Accordingly, as the taxpaying labour force contracts as a proportion of the total population, with many at the top able to opt out of paying their share, far more revenue will be needed for schools, universities, hospitals, infrastructure (including submarines), and wages for teachers, doctors, nurses, soldiers, teachers, academics, and researchers.

The use of trusts registered in tax havens such as the Cayman Islands means that for many of the super rich, paying taxes is essentially voluntary. Workers at McDonald's rarely use trusts or tax havens.

Work can be classified as 'tangible' or 'intangible'. Physical activity in a factory, department store, warehouse, school, hospital, or restaurant is easy to quantify and to tax—usually with taxation deducted at the source. But intangible work in a digital economy, processing symbols—words, sounds, images, numbers, and designs—often across oceans and continents, is extremely difficult to measure. About 40 per cent of Australians now work in the 'information' sector, broadly defined.

Australia's total tax take is at the lower end of the OECD average, but collecting additional revenue from income may be politically impossible.

The Goods and Services Tax is a major source of revenue, essentially for the states, but, as a flat tax of 10 per cent, is regressive, taking a higher percentage of income from the poor than from the rich. It is easy to collect at the point of purchase, and, twenty years since its introduction, there are signs that the federal and state governments want to minimise exclusions to it and increase its rate.

Conversely, offering income-tax cuts, even when more revenue is needed, has proved to be a political winner for governments of all persuasions, based on the premise that lower taxation will generate higher levels of investment, more growth, and, as the cliché goes, 'A rising tide lifts all boats.' But it will do nothing to reduce the growing inequality in Australia. Money is seen as an incentive for the rich and a disincentive for the poor: the Coalition promotes the dogma that putting money in the pockets of the affluent stimulates investment and economic activity, while it makes the poor lethargic and welfare dependent ('Galbraith's law').

An estimated 1.3 million families, many in the middle-income bracket, have built their financial security by using taxation anomalies such as negative gearing, whereby paying off additional properties can be offset against income and franking credits. Challenging this—as the ALP found in the 2019 election—comes with a heavy price.

The political sensitivity of taxation means that it is never rationally debated. Its profound moral implications—balancing immediate gains against long-term security—are ignored or discounted, and this has become a major factor in retarding our effective response to climate change and transitioning to a post-carbon economy.

Moral choices

Lord Acton, the great English Catholic liberal historian, wrote some compelling sentences about moral factors in history:

I exhort you never to debase the moral currency or lower the standard of rectitude, but to try others by the final maxim that governs your own lives, and to suffer no man and no cause to escape the undying penalty which history has the power to inflict on wrong ...

[The progress of civilization] depends on preserving, at infinite cost, which is infinite loss, the crippled child and the victim of accident, the idiot and the madman, the pauper and the culprit, the old and infirm, curable and incurable. This growing dominion of disinterested motive, this liberality towards the weak, in social life, corresponds to that respect for the minority, in political life, which is the essence of freedom.[2]

We are individually and collectively faced with moral choices every day. Philosophers and ethicists agonise over the moral dilemma involved when we know that acting to ensure a personal or national benefit will inflict loss and destruction on others. How should we choose?

As the philosopher Tony Coady puts it:

We are not faced with choosing whether to kill one entity to save many, or one large number of people to save another larger number, or one nation to save many. It's more that we might have to choose some privations, make some sacrifices, in order to save many lives (and environments). The choice of restrictions now on comfortable living and some wellbeing in order to avoid disasters to everyone in the (probably relatively near) future is not like violating a deep moral obligation (like the prohibition of intentional killing of an innocent person) for a single case in order to save the killing of some number more.[3]

Australia faces an ethical challenge about action on climate change, following the COVID-19 lockdown and the need, we are told, for a rapid return to normality.

Australia maintains the highest priority for our fossil-fuel exports, offering the drug-dealers' defence. ('If we don't sell it, someone else will.') We can't help ourselves.

If Australia maximises its exports of coal and natural gas to China and India, it would make a major contribution to GDP and to employment—even to tax revenue, if all the mining corporations could be induced to pay. But if this made a progression towards a 2-degree increase in mean global temperatures irreversible, with a sharp increase in the frequency of extreme weather events— including bushfires—then the collective memory of our self-discipline over a period of months to address a pandemic would seem to have been a trivial sacrifice compared with the threat to the planet, its inhabitants, and the biota.

But to change our public policies and our view (if any) of our role as global citizens, it appears that we need a new narrative; facts won't do it on their own. Oddly, in the 1970s, Gough Whitlam and South Australian Labor premier Don Dunstan, to name two outstanding examples, could explain very complex and often controversial issues— in Whitlam's case, in speeches of heroic length—and win public support. Hawke and Keating had a similar gift. Who has it now?

Always take the long view

Psychologically, we react instinctively to immediate threats or opportunities. Our species beat the Neanderthals because our responses were (presumably) quicker. But then modern humans began planning ahead for the next season, planted crops, and were no longer exclusively hunter-gatherers or grazers.

The year 2030 has to be planned for now, and we must work towards a global network to forestall the potential disaster of 2050.

Global emissions must be cut by 7.6 per cent for each year of this decade. The greatest challenge for us all is to enable humanity to achieve its full potential, not just as consumers, and to preserve our home, our planet, to understand what we are capable of.

It is, I suspect, not an accident that the study of humanities at universities has been been singled out for discrimination by sharply increasing charges in a sector already badly damaged by COVID-19 and the withdrawal of overseas fee-paying students.

We must resist the smug and dangerous implication: Who needs philosophers, historians, political scientists, psychologists, journalists, critics, anthropologists, archaeologists, writers, musicians, and creative artists, just because they can throw light on the human condition and help us to find out who we are?

We cannot be part-time humans.

The huge task of exploring human potential has never been taken seriously. Nor has the equally huge task of meeting human needs.

It is about time we found out.

A Gettysburg Address for 2020

Securing popular support for courageous changes—many of them painful in the short term—will require developing an appropriate narrative that can persuade and unite. Great leaders have that gift.

In 1860, Abraham Lincoln became the first Republican to be elected as president of the United States. Lincoln was reflective and self-doubting, and spoke in evidence-based testable propositions, using sentences that contained verbs. He appealed to 'the better angels of our nature'. He never used his own name in a speech, clapped himself, or wore a baseball cap. He wrote wonderful letters.

On 19 November 1863, at Gettysburg, Pennsylvania, Lincoln delivered a famous short address, just 271 words long, at the inauguration of a war cemetery, in which he defined the nature of democracy in the context of a brutal civil war over slavery.

I have often speculated about what Lincoln might have said if he had been addressing contemporary issues. I have attempted a version of this and my draft is exactly the same length. There are fifteen echoes of Lincoln's text in mine: the words in inverted commas are Margaret Thatcher's, from 1988:

A score of years ago, we entered a new millennium, facing great challenges. World population explodes; in both rich and poor nations men and women live far longer, and now consume the Earth.

Earth's raw materials are finite. Water, forests, farmlands are threatened by 'a massive experiment with the system of the planet itself', causing climate change, droughts, floods, hurricanes. Rich, powerful nations exploit the weak and paralysed.

Now we are engaged in great global conflicts of values. Gaps between inconceivable wealth and desperate poverty create hatred, wars, fundamentalism, and terrorism.

Science and technology destroy boundaries, but nations turn inward, becoming tribal; political leaders reject global goals of compassion, reconciliation, and understanding. Racism, nationalism, militarism, religious hatred, democratic populism, suppression of dissent poison democracy's wells. Some leaders use propaganda, resolve problems by suppression, promote fear of difference, attack organised labour, weaken the rule of law, use state violence, torture, execution. Evidence-based policies are displaced by appeals to fear and anger.

The great tasks before us are to dedicate ourselves to overcoming fear of difference, recognise that environment and economy are bound together. The human condition is fragile, and we must not confuse prejudice with principle.

We must consecrate ourselves to the unfinished work of saving Planet Earth, our home, where our species, *Homo sapiens*, lives and depends for survival. All nations, and all people, must dedicate themselves to protecting our global home instead of promoting national, regional, or tribal interests. We must highly resolve to save the air, save the soil, save the oceans to ensure that our species, and the noblest aspects of its culture, shall not perish from the Earth.

It is essential that we not fall into despair, and retreat to the caves. But citizens have to be informed, and then challenge and speak truth to power. It will not be easy. It will be exhausting. It will not be comfortable. We will probably lose some friends. But it must be done.

Acknowledgements

I would like to express my grateful thanks to Peter Doherty, John Zillman, Mike Manton, Ian Enting, Ross Garnaut, Elizabeth Truswell, David Shearman, Sarah Brenan, Nan McNab, Louise Sweetland, Rachel Faggetter, Paul Kelly, Peter Colman, Peter Singer, Richard Flanagan, Tony Coady, and Phillip Adams.

Henry Rosenbloom is not only the publisher of this work, but also its tireless editor, through many iterations. His contribution has been extraordinary, combining drive, passion, erudition, sensitivity, and the capacity to anticipate and solve problems. I am very much in his debt.

Mike Richards was also a valued counsellor, fact-checker, and co-editor.

Notes

Chapter One: Where We Begin: *Sleepers, Wake!* in 1982

1 Victorian Parliamentary Debates, 12 September 1972, vol. 308, p. 195; Hansard, House of Representatives, 23 February 1978, vol. 108, p. 164.

2 The other members nominated by Beazley's office were Mairéad Brown, Jane Marceau, Bob Williamson, Cathy Zoi, John McFarlane, Clem Doherty, Russell Bate, John Boyd, Peter Dixon, Rosemary Herceg, Kate Lundy, Aidan McCarthy, and Nabeel Youakim .

Chapter Two: Democracy's Existential Crisis in a Post-truth Era

1 I identified the West's long history of failed interventions in the Middle East ('Middle Eastern Horrors') in *Knowledge Courage Leadership* (Wilkinson Publishing, 2016), p. 254.

Chapter Three: Overturning the Enlightenment

1 Yuval Noah Harari, *21 Lessons for the 21st Century* (Random House, 2018), p. 302.

Chapter Four: How the Digital World Changed Everything

1 Sir Tim Berners-Lee (1955–) studied physics at Oxford, holds a chair at MIT, is a professorial fellow at Oxford, chairs the World Wide Web Consortium (W3C), was elected a Fellow of the Royal Society, and received a knighthood (KBE) and the Order of Merit (OM).

2 'Can Mark Zuckerberg Fix Facebook before it Breaks Democracy?' by Evan Osnos, *New Yorker*, 17 September 2018.

3 'The Autocracy App', by Jacob Weisberg, *New York Review of Books*, 25 October 2018.

4 In 1993, General Motors had been by far the world's largest corporation, employing 710,000 workers. By 2020, it ranked no. 13 in the US with 170,000 employees, with a relatively modest market capitalisation of $US40 billion, and was no. 4 in the world as a motor manufacturer.

5 Claude Shannon (1916–2001) worked at the Bell Labs from 1941, exchanged ideas with Alan Turing, then became a professor at MIT in 1956. With Warren Weaver, he wrote the definitive text on information theory, *The Mathematical Theory of Communication* (1949). When I visited him in 1981, he played ping-pong with a robot (automaton) he had built.

6 Turing helped design the ACE and MADAM computers in Manchester after the war. Stigmatised as a homosexual, he committed suicide after conviction for an offence. Posthumously pardoned in 2013, his face appears on the £50 banknote, and he is regarded as an authentic hero (and victim) of our time. His wide range of interests included artificial intelligence and mathematical biology. His work on morphogenesis (how organisms develop their shape) anticipated research on the structure of DNA.

   During World War II, Turing worked as a code-breaker at Bletchley Park, and broke the German Enigma naval code, first with a smaller machine, oddly called 'the bombe', which he adapted from a Polish prototype, but later contributed to Colossus, the world's first programmable electronic computer. He is credited with helping to win the Battle of the Atlantic and shortening the war, perhaps by up to two years, but was miserably rewarded with an OBE in 1945.

7 Adrienne Lafrance, *The Atlantic* (June 2020), 'Nothing Can Stop What is Coming' (https://www.theatlantic.com/magazine/archive/2020/06/qanon-nothing-can-stop-what-is-coming/610567/).

Chapter Five: The Trump Phenomenon

1 *New York Review of Books*, 27 February 2020.
2 Simon Schama, 'Liberalism, Populism and the Fate of the World', The 2018 Leslie Stephen Lecture, Cambridge.
3 *New York Review of Books*, 6 June 2019.

Chapter Six: Climate Change: the science

1 Lavoisier made a fortune as a 'tax farmer'—collecting revenue for the state—and used it to set up an elaborate experimental laboratory. He also made an enemy of Jean-Paul Marat. A victim of the Terror during the French Revolution, he and his father-in-law were guillotined in 1794 for being landlords, not scientists. Over two centuries later, in a supreme act of cynicism, a group of Australian climate-change denialists adopted the name of the Lavoisier Group.
2 The full official title of the report, which is known as SR15, is *Global warming of 1.5°C: an IPCC special report on the impacts of global warming of 1.5°C above pre-industrial levels and related global greenhouse gas emission pathways, in the context of strengthening the global response to the threat of climate change, sustainable development, and efforts to eradicate poverty.*
3 Crutzen, P.J., & Stoermer, E.F., *The Anthropocene. International Geosphere–Biosphere Programme.* Newsletter 41 (2000).

Chapter Seven: Climate Change: the politics

1 These figures come from energy.gov.au, and apply to 2018.
2 'Pascal's wager' was first proposed by the seventeenth-century French philosopher Blaise Pascal in considering the existence of God, and is found in Part III, § 233, of his *Pensées* (Penguin).

Chapter Eight: Retail Politics: targeted, toxic, trivial, and disengaged

1 A.C. Grayling, *Democracy and Its Crisis* (Oneworld Publications, 2017), p. 190.

Chapter Nine: The Death of Debate: the loss of language and memory

1 I drop in the names hesitantly because all but a few will be unknown to anyone aged less than 60: Gough Whitlam, Paul Keating, William McMahon, Malcolm Fraser, John Howard, Fred Chaney, Peter Nixon, David Thomson, Ian Sinclair, John Carrick, Bill Yates, Clyde Cameron, Ralph Jacobi, Mick Young, Dick Klugman, Jim McClelland, John Wheeldon, Neal Blewett, Gareth Evans, Susan Ryan, John Button, John Kerin, Brian Howe, Peter Baume, Ian Macphee, Ian Viner, Bob Ellicott, Jim Carlton, Margaret Guilfoyle, Alan Missen, Ralph Willis, Clyde Holding, Don Grimes, Moss Cass, Janine Haines, Tom Uren, Jim Killen, and Charles Jones, and I might modestly add myself. Most can be tracked down on Wikipedia, and I commend the exercise.

Chapter Eleven: Being Honest with Ourselves

1 *The Wayward Tourist: Mark Twain's adventures in Australia.* Introduction by Don Watson (Miegunyah Press, 2007).

2 Peter Cochrane, *Best We Forget: the war for White Australia, 1914–18* (Text, 2018).

3 Thomas Piketty, *Capital and Ideology* (Belknap/Harvard, 2020), p. 526.

4 Op. cit., p. 813.

5 Op. cit., p. 841.

Chapter Twelve: The Corona Revolution

1 The World Wealth Organization renamed it 'COVID-19' [Corona virus disease-2019] on 11 February 2020.

Chapter Thirteen: Saving the Planet

1 Simon Schama, 'Liberalism, Populism and the Fate of the World', The 2018 Leslie Stephen Lecture, Cambridge.

2 WHO Health and Climate Change Survey Report, 2019.

Chapter Fourteen: What Is to Be Done: political engagement and climate change

1   Ross Garnaut, *Superpower: Australia's low-carbon opportunity* (La Trobe University Press, 2019).

2   *Introductory Lecture on Modern History* (1895), John Emerich Edward Dalberg Acton, 1st Baron Acton (https://oll.libertyfund.org/titles/acton-lectures-on-modern-history/).

3   Private correspondence.

Bibliography

Books

Acton, 1st Baron *Introductory Lecture on Modern History* (1895) (https://oll.libertyfund.org/titles/acton-lectures-on-modern-history)

Albright, Madeline *Fascism: a warning* (Harper Collins, 2018)

Andrew, Christopher *The Secret World* (Penguin Books, 2018)

Arendt, Hannah *The Origins of Totalitarianism* (Penguin Books, 1951)

Beeson, Geoff *A Water Story: learning from the past, planning for the future* (CSIRO Publishing, 2020)

Bell, Daniel *The Coming of Post-Industrial Society: a venture in social forecasting* (Basic Books, 1973)

Brett, Judith *The Coal Curse: resources, vlimate and Australia's future* (Quarterly Essay #78, Black Inc., 2020)

Butler, Mark *Climate Wars* (Melbourne University Press, 2017)

Cameron, Sarah and McAllister, Ian *Trends in Australian Political Opinion: results from the Australian Election Study 1987–2019* (Australian National University, 2019)

Cochrane, Peter *Best We Forget: the war for White Australia 1914–18* (Text, 2018)

CSIRO *Climate Change: science and solutions for Australia* edited by Helen Cleugh, Mark Stafford Smith, Michael Battaglia, and Paul Graham (CSIRO Publishing, 2020)

Davies, William *Nervous States: how feeling took over the world* (Norton, 2018)

Davis, Mark *The Land of Plenty: Australia in the 2000s* (Melbourne University Press, 2008)

Dawkins, Richard *The Selfish Gene* (Oxford University Press, 1976)

Diamond, Jared *Upheaval. how nations cope with crisis and change* (Allen Lane, 2019)

Doherty, Peter *Pandemics* (Oxford University Press, 2013)

Ehrlich, Paul and Howland, Anne *The Population Bomb* (Ballantine Books, 1968)

Fisk, Robert *The Great War for Civilisation: the conquest of the Middle East* (Fourth Estate, 2005)

Fukuyama, Francis *The End of History and the Last Man* (Penguin Books, 1992)

Garnaut, Ross *The Garnaut Climate Change Review.* Final Report (Cambridge University Press, 2008)

—*Superpower: Australia's low-carbon opportunity* (Black Inc., 2019)

Grayling, A.C. *Democracy and Its Crisis* (Simon & Schuster, 2017)

—*The History of Philosophy* (Penguin Books, 2019)

Harari, Yuval Noah *21 Lessons for the 21st Century* (Jonathan Cape, 2018)

Horne, Donald *The Lucky Country* (Penguin Books, 1964)

Huxley, Aldous *Brave New World* (Penguin Books, 1932)

Isaacson, Walter *The Innovators* (Simon & Schuster, 2014)

Jensen, Erik *The Prosperity Gospel* (Quarterly Essay #74, Black Inc., 2019)

Jones, Barry *Sleepers, Wake! technology and the future of work* (Oxford University Press, 1982; 4th edition, 1995)

—*Knowledge Courage Leadership* (Wilkinson Publishing, 2016)

Judt, Tony *Ill Fares the Land* (Penguin Books, 2011)

Kahneman, Daniel *Thinking Fast and Slow* (Penguin Books, 2012)

Kaufmann, Eric *Whiteshift: populism, immigration and the future of white minorities* (Penguin Books, 2018)

Keane, John *The Life and Death of Democracy* (Simon & Schuster, 2009)

—*The New Despotism* (Harvard, 2020)

Kelly, Dominic *Political Troglodytes and Economic Lunatics* (Black Inc., 2019)

Lakoff, George *Don't Think of an Elephant: know your values and frame the debate* (Scribe, 2004)

Leigh, Andrew *Battlers & Billionaires: the story of inequality in Australia* (Redback, 2013)

—*Choosing Openness* (Penguin Books, 2017)

Levitsky, Steven and Ziblatt, Daniel *How Democracies Die: what history reveals about our future* (Penguin Books, 2019)

Lewis, Michael *The Fifth Risk* (Allen Lane, 2019)

Lewis, Peter *Webtopia: the world wide wreck of tech and how to make the net work* (NewSouth Publishing, 2020)

Lynch, Jack *You Could Look It Up: the reference shelf from ancient Babylon to Wikipedia* (Bloomsbury, 2016)

Marmot, Michael *The Health Gap: the challenge of an unequal world* (Bloomsbury, 2015)

—*Fair Australia: social justice and the health gap* (ABC, Boyer Lectures, 2016)

Marx, Karl *Grundrisse: foundations of the critique of political economy* (translated by Martin Nicolaus, Penguin Classics, 1993)

Mayer, Henry and Nelson, Helen (eds) *Australian Politics: a fifth reader* (Nelson, 1980)

McKenna, Mark *Moment of Truth: history and Australia's future* (Quarterly Essay #69, Black Inc., 2018)

McKibben, Bill *The End of Nature* (Anchor, 1989)

— *Falter: has the human game begun to play itself out?* (Black Inc., 2019)

McLuhan, Marshall *The Medium is the Massage: an inventory of effects* (Penguin Books, 1967)

—*War and Peace in the Global Village* (Bantam, 1968)

McMichael, Anthony J. *Human Frontiers, Environments and Disease* (Cambridge University Press, 2001)

McMichael, Anthony J, Woodward, Alistair, and Muir, Cameron *Climate Change and the Health of Nations: famines, fevers, and the fate of populations* (Oxford University Press, 2017)

Mishra, Pankaj *Age of Anger: a history of the present* (Allen Lane, 2017)

Myers, Sir Rupert, et al., *Technological Change in Australia.* Report of the Committee of Inquiry into Technological Change in Australia (CITCA) (4 vols, Australian Government Publishing Service, 1980)

Orwell, George 'Politics and the English Language' (1946), in *Essays* (Penguin Modern Classics, 2000)

—*Nineteen Eighty-four* (Penguin Books, 1949)

Pearse, Guy *High and Dry: John Howard, climate change and the selling of Australia's future* (Penguin Books, 2007)

Piketty, Thomas *Capital and Ideology* (translated by Arthur Goldhammer; Belknap/Harvard, 2020)

Pilling, David *The Growth Delusion: the wealth and well-being of nations* (Bloomsbury, 2018)

Pinker, Steven *Enlightenment Now: the case for reason, science, humanism and progress* (Penguin Books, 2018)

Postman, Neil *Amusing Ourselves to Death: public discourse in the age of show business* (Viking Penguin, 1985)

Roth, Philip *The Plot Against America* (Vintage, 2004)

Runciman, David *How Democracy Ends* (Profile, 2018)

Rusbridger, Alan *Breaking News: the remaking of journalism and why it matters now* (Canongate Books, 2018)

Sachs, Jeffrey *The Price of Civilization: economics and ethics after the fall* (Bodley Head, 2011)

Schama, Simon 'Liberalism, Populism and the Fate of the World', the 2018 Leslie Stephen Lecture, Cambridge, in *Wordy: sounding off on high art, low appetite and the power of memory* (Simon & Schuster, 2019)

Snyder, Timothy *The Road to Unfreedom: Russia, Europe, America* (Jonathan Cape, 2018)

—*On Tyranny: twenty lessons from the twentieth century* (Bodley Head, 2017)

Soutphommasane, Tim *On Hate* (Hachette, 2019)

Stephens, David and Broinowski, Alison (eds) *The Honest History Book* (NewSouth Publishing, 2017)

Stiglitz, Joseph *The Price of Inequality* (Norton, 2012)

—*People, Power and Profits* (Penguin Books, 2020)

Thompson, Mark *Enough Said: what's gone wrong with the language of politics* (Bodley Head, 2016)

Turnbull, Malcolm *A Bigger Picture* (Hardie Grant, 2020)

Twain, Mark / Don Watson *The Wayward Tourist. Mark Twain's adventures in Australia* (Miegunyah Press, 2007)

Watson, Peter *The German Genius* (Harper Perennial, 2010)

Wheen, Francis *Marx's Das Kapital: a biography* (Allen and Unwin, 2006)

Zuboff, Shoshana *The Age of Surveillance Capitalism: the fight for a human future at the new frontier of power* (Hachette, 2019)

Reports

Managing Climate Change. Papers from the Greenhouse 2009 Conference. Imogen Jubb, Wenju Cai, and Paul Holper (eds) (CSIRO Publishing, 2010)

House of Representatives Standing Committee for Long Term Strategies 'Australia's Population "Carrying Capacity": one nation—two ecologies' (Australian Government Publishing Service, 1994)

Knowledge Nation Taskforce *An Agenda for a Knowledge Nation* (Chifley Research Centre, 2001)

Index